Global Technography

Intersections in Communications and Culture

Global Approaches and Transdisciplinary Perspectives

Cameron McCarthy and Angharad N. Valdivia
General Editors

Vol. 24

PETER LANG
New York • Washington, D.C./Baltimore • Bern
Frankfurt am Main • Berlin • Brussels • Vienna • Oxford

Grant Kien

Global Technography

Ethnography in the Age of Mobility

PETER LANG
New York • Washington, D.C./Baltimore • Bern
Frankfurt am Main • Berlin • Brussels • Vienna • Oxford

Library of Congress Cataloging-in-Publication Data

Kien, Grant.
Global technography: ethnography in the age of mobility /
Grant Kien.
p. cm. — (Intersections in communications and culture; v. 24)
Includes bibliographical references and index.
1. Communication in ethnology. 2. Communication and culture.
3. Culture and globalization. 4. Technological innovations.
5. Mobile communication systems. 6. Mobile computing. I. Title.
GN307.5.K54 306—dc22 2008053503
ISBN 978-1-4331-0294-3 (hardcover)
ISBN 978-1-4331-0293-6 (paperback)
ISSN 1528-610X

Bibliographic information published by **Die Deutsche Bibliothek**.
Die Deutsche Bibliothek lists this publication in the "Deutsche
Nationalbibliografie"; detailed bibliographic data is available
on the Internet at http://dnb.ddb.de/.

Book interior designed by Kevin Dolan

The paper in this book meets the guidelines for permanence and durability
of the Committee on Production Guidelines for Book Longevity
of the Council of Library Resources.

Printed in the United States of America

Contents

Illustrations vii

Acknowledgments ix

INTRODUCTION
Mobility, Being, Global Network 1

CHAPTER ONE
Global Technography: Ethnography in the Mobilized Field 9

CHAPTER TWO
Network and Power: The Global Landscape 27

CHAPTER THREE
Technological Mobility and Cultural Practice 57

CHAPTER FOUR
Ontology, Technological Mobility, and "Belonging" 71

CHAPTER FIVE
Mobility, Globalization, and Culture 85

CHAPTER SIX
Post-Global Citizenship 103

CHAPTER SEVEN
A Theory of Home for the Mobile, Globalized Citizen 119

CHAPTER EIGHT
Technology Is Human 137

APPENDIX
From Heidegger to Technography: A Way Outward in a Distanceless World 147

Glossary of Actor-Network Theory Terms 169

Bibliography 179

Index 191

Illustrations

FIGURE 1-1
A View of Itaewon from the Balcony
of Star Express 11

FIGURE 1-2
Laptop Pirates on Hou Hai Lake 25

FIGURE 2-1
First Street and Springfield Avenue
in Champaign, Illinois 28

FIGURE 3-1
Nationalism Online 58

FIGURE 4-1
Swami Nitya Muktananda in His Yoga Studio 75

FIGURE 5-1
Children of Chiapas 91

FIGURE 6-1
CBC Online 110

FIGURE 7-1
Toronto Icons 120

Acknowledgments

Thank you to Cameron McCarthy, Norman Denzin, Clifford Christians, and James Hay. Your assurances allowed me to find the courage to follow my own intellect, explore the unconventional, and invent what has been needed to bridge the spaces between. Thanks to Jia Jia for having confidence in me. Thanks to Sean Nevins, a rare friend. Thanks to Swami Nitya Muktananda, my patient teacher, friend, and comrade in headstands and meditation. Thanks to my family. Big thanks to Kevin Dolan for sharing your technical expertise. Thanks to Sharon Tettegah, Paula Treichler, my colleagues James Salvo and Michael Giardina, my guides Yoshitaka and Koji, Father John, Zhou Yuanzhi, the staff of Hoyah Academy, Yan Rui, the staff and guests of the Juyoh Hotel, los hermanos Baldizón, Veronica, Walter Podilchack, Luz Briseida, the people of La Nueva Revolución Chiapas, Michael Kilcullen, Celiany Rivera-Velázquez, Ted Gournelos, Shoshana Magnet, and all the unknown lovers and incognito citizens of our global village. And of course, the CBC—one could hardly be Canadian without you! Thanks to the Graduate College of the University of Illinois at Urbana-Champaign and the International Institute of Qualitative Inquiry for financial support. This research would not have been possible without support from the Social Science and Humanities Research Council of Canada, so a BIG thank you to my fellow Canadians!

INTRODUCTION

Mobility, Being, Global Network

This book develops and exemplifies an ethnographic methodology I am calling Global Technography.[1] I was inspired to innovate this methodology in due course while exploring the broad social phenomenon of mobility in the context of global network, media, and everyday cultural performances in which human actors and media technologies interact to produce the meaningful encounters that we know as our life experiences. Finding the tradition of conceptually demarcating and enclosing a space designated as the ethnographic field to be untenable in documenting how mobility is actually experienced in everyday life, I instead work with the notion of momentary spatialization as a means of finding and defining a physical location for observation. In addition to this methodological innovation, two further aspirations have driven this work forward. First, I offer an example of how sociotechno researchers, scholars in science and technology studies, philosophers of technology, and qualitative researchers can escape the traditional interactionist approach to technology that treats devices and machinery as dead props. Rather, I seek to illustrate how technology dynamically works with human actors to create and maintain the world we live in. Secondly, I aspire to demonstrate to researchers and scholars in general that acknowledging and documenting the participatory role of technology in everyday life and culture need neither sensationally celebrate technology nor denounce it out of paranoia. Instead, I wish to make obvious the value of exploring the role of technology in a way consistent with how we actually experience everyday life. A quick review of how I got to this place is in order.

The year was 1997 in Boboli Gardens, Florence, Italy. Like birds calling throughout the gardens, cell phone ringers would randomly break the silence from all directions, prompting their owners to squawk their singular reply, "Pronto!" It created a cacophony of precise technological noise mixed with the absurdity of human communication. Then, like a shot breaking through the afternoon air, I heard a crisp, English "Hello." I suddenly realized the significance of the wireless future of communica-

1 For an overview of technography, see Kien (2008).

tions technology. I could see that much of the wired infrastructure could be bypassed altogether, that the advance of personal appliances would make mobility the new normal circumstance on a global scale. I realized that "wireless" means everywhere, anywhere, any moment, every moment. Like others, I had been aware of the Zapatista uprising in Chiapas, Mexico, which was said to be possible in large part by the globalization of wireless communication that allowed them to globalize their brand of resistance, in effect using the tools of neoliberalism against themselves. As a result, the Zapatista's struggle was almost as much about communication and aesthetics and turning the global network on itself as it was about confrontation in physical space. However, the full impact of what the mobile evolution in technology meant in everyday life didn't hit me until that day in Boboli Gardens. It was my moment of epiphany, after which my understanding of globalization and communications would never be the same again. There would no longer be a need to go to technology, or even for technology to come to you—once networked successfully, it becomes part of our ontology and we are simply never apart from it again. Although mobility and human influence on environment are not new phenomena, we are living through a seismic shift in our media environment that is bringing with it unprecedented freedoms and changes in power relationships that are enacted in the physical world through mobility. It's already banal to point out how the array of products on and coming to the market demonstrate that technology is rapidly progressing from portability to extreme mobility, to be used reliably while in transit—to be part of the experience of transit—making constant movement and positional fluidity to be a normal condition of technological subjectivity. Meanwhile, transit itself is a new normal. There is a class of global citizen constantly in motion for whom the term "migration" is meaningless, as the act of settlement isn't much part of their everyday experience. This is the set of circumstances in which evolve the portrait of the technologically mobile subject: a hypermediated 24/7 virtual environment (Kien, in press a). Advanced miniaturized technology, especially in medicine,[2] is designed to function both inside and outside, to travel with and within the human body as both enhancement

2 With approval from the United States Food and Drug Administration (see http://www.fda.gov/cdrh/emc/wmt-about.html) and the U.S. Federal Communications Commission (see http://wireless.fcc.gov/services/personal/medtelemetry/). For a downloadable study guide on wireless medical technology development, see http://www.monitoring.welchallyn.com/products/wireless/resourcelib.asp.

and appendage. Historical precedents given their due,[3] it is the networked nature of new digital appliances that distinguishes them from their predecessors. National territory is no longer confined to the physical borders of the nation, it rather erupts in the spontaneous performances of citizens in transit. As an ethnographic project, the field itself needs redefinition to accurately reflect the nature of space in this age of global mobility.

While theory and philosophy are strategically deployed throughout this book, it is my intention that vignettes illustrating eruptions of everyday experience should form the basis of a conversation between idealistic theoretical absolutes and the messiness of life on the ground as we live it. The work presented in this book takes a cue from Carey's directive, "to examine the actual social process wherein significant symbolic forms are created, apprehended, and used" (Carey, 1989, p. 30). My interpretation of Carey's words is informed by Heidegger's philosophy of technology, in which technologies are created to be intrinsically part of ontological experience. In other words, they are created to perform roles—to act—in everyday life, and this demands an ethnography that can track the social performativity of technology as well as human beings. The performance of technology in everyday life is an important aspect informing the invention of Actor-Network Theory (ANT), which offers a methodological technique in which dichotomies such as "network" versus "individual" can be reconciled.[4] However, the technical details of technical relationships can sometimes be lost or simply ignored in the exploration of performative moments. When there are neither individuals nor power, but only "collectifs" (Callon and Law, 1995) and "network effects" (Latour, 1986) in which singularity is an aesthetic slight of hand,[5] the experience of how one actually tends to conduct oneself and be treated—as an individual singular entity—can feel lacking. Theoretical interventions can help reconcile the nuances of individual performativity (of both human and nonhuman actors) within schematized geographical and geometric concepts, while instances drawn from messy everyday situations make such speculations understandable in life as we experience it. Practice constantly exceeds the field and any theoretical attempt to define it. Everyday life is complex and

3 I.e., walkie-talkies, transistor radios, the walkman.

4 For an overview of my interpretation of ANT, see Kien in press b.

5 For an ANT definition of "collectifs," see Glossary, p. 170; for "network," p. 173; for "singularity," p. 175.

confusing. In the emerging tradition of critical qualitative inquiry, I resist reducing everyday experiences to one-dimensional anecdotes in support of quasi post-positivist theorizing. Rather, I assume it is the everyday that stands on its own, without need of justification. It is the speculation we call theory that grasps desperately at threads of legitimation for its relevance to lived reality, for it is us as actors who seek meaning and justification for our actions.

Truth can be found only through technology, since truth itself is ideal and needs to be demonstrated, brought forth, and made to present itself. Benjamin (1978) gave us the terminology of the "optical unconsciousness" to describe how we believe we cannot know truth without technological assistance, and of "technological loyalty" to describe how we come to rely on machines to show us the truth about the universe we live in. Heidegger also embraced this idea, explaining that we need technology to manufacture situations in which the truth may erupt and become knowable. These eruptions then come to inform us about the truth of our existence. Self-inquiry, intellectual methodologies, ceremony, and ritual are ways of bringing forth the art of the self, the poiesis of being a person. Heidegger's postulate is heightened by technologically mobile media. The cell phone and laptop, now converging in the smartphone, are standard, functional pieces of the global citizen's mobile uniform. My use of the term mobility refers to the movement of individual people and actants in ever-shifting arrays that comprise actor-networks: not only movements in physical terms, but also conceptual movements such as transience in identity. Being refers to a Heideggerian state of "pure being," moments of revelation in which the truth about meaning and how we enframe our worlds erupt. Throughout this book, I work with this as "epiphany" in the Denzinian school of interpretive inquiry, a turning point after which, having gained "authentic understanding" (Denzin, 2001), things are no longer the same. Finally, network refers to the ever-mutating actor-network for which the global international telecom system serves as the most stabilizing entity. This entity is always participating in and allied with other, often more dominant networks of translation. For example, through its regulatory apparatus, nations partition the network and claim parts of it as their own. No less, nations take advantage of the network to extend themselves within and beyond their demarcated national borders, translating mediated space into nationalist territory, implicating communications technology in ritualized performances of culture.

Global Technography as I am defining it is interdisciplinary, involving performance studies, ethnography, Actor-Network Theory, media effects theory, globalization and identity theory, and some applied technological savvy. My wish is to contribute another step towards "a fundamentally new direction...in the approach to culture and identity" involving "a radical rethinking of the linkages of knowledge, culture, and association among people" (McCarthy et al., 2003, p. 30). Seippel (2008) suggests a network analysis is appropriate in understanding social behavior for three important reasons:

1. It captures what most of us actually see as a fundamental aspect of social life ontologically: the relatedness and/or embeddedness of human beings and social interactions.
2. The most vital characteristics of social interaction within civil society and civil society's autonomy in relation to other social institutions hinge on relations and networks, and networking is vital to social capital.
3. The relational character of power and influence are best understood as network formations.

Wilson (2007) argues for a need to employ ethnography on a global scale in order to understand relations of domination and resistance.

To state my own theoretical and philosophical assumptions, I acknowledge that for me Heidegger's philosophy of technology and his phenomenology are a satisfactory description of the nature and relationship of technology and my own existence. I believe Heidegger is correct in suggesting that technology cannot be separated from ontology since it is profoundly part of human everyday experience, emphatically so if globalization theorists are to be taken seriously. Problematic as it might be, I accept that Leibniz's monism—the philosophical basis of digital code (see Kien, 2002)—is the description of the nature of existential force that our digital world unspokenly assumes, which prompts reconciling dualism's theoretical inside/outside essentialisms and dichotomies. I work from the media studies perspective that media has the potential to profoundly affect culture and yet have enough of the cultural studies scholar in me to consider the relationship dialectical in that culture often determines the ways technologies are put to use. I assume everyday life is comprised of mundane, subtle, and sometimes spectacular performances, and all are equally important in the revelation of truth. I consider Actor-Network Theory a compliment to Heidegger's philosophy of technology, providing a pragmatic theoretical grounding that can describe mobility and the relationship of "network and site" (Hay, 2001, p. 213).

While overcoming dualism is one attractive feature of ANT, it also emphasizes ethnography as a method, interrogating everyday life as semiotic performance and effects of performances. This keeps pace with developments in autoethnography, although much qualitative research continues to reify human beings as the only source of agency and thereby to reify dualisms, as Durkheim and Goffman-rooted ethnography tends to treat technology as object and prop outside of meaningful human interaction. With ANT as a theoretical grounding for the ethnographic study of wireless digital technology, I focus on the processes and effects of global ontology that are enabled by technology. I undertake to demonstrate some of the mechanics of how some technologies actually do social work and to elucidate moral effects and rehearsals of actual and preferred outcomes (Denzin, 2003). Overall, my work shows some of the alliances, betrayals, struggles, and triumphs that go towards the construction of identity (including nationalism and hybridity) and performances of self in the context of globalization.

This book begins with a discussion of the methodology invented for this research: Global Technography. The issue of territorial definition is discussed in the context of mobility, with a rethinking of the traditional method of defining a closed field for observation resulting from the unique circumstances of technologically enabled mobility. Spatializing eruptions of mobile technology use—technological performativity—are taken as the defining feature of the field in the form of global network. Numerous vignettes exemplify the characteristics of such performative moments and their complex interconnectedness with everyday concerns.

Chapter 2 deals with the subject of global network and power, exploring various theoretical descriptions of how power operates in the context of complex international exchanges. Site-specific vignettes help understand how power is experienced in everyday "global" situations, inflected with highly nuanced histories, cultural logics, and intensely personal motivations. Chapter 3 then deals more squarely with the relationship between cultural logic and the use of mobile technology. This chapter pointedly explores the performance of mobility as cultural practice, illustrating how the way one uses technology may be profoundly informed and directed by cultural values.

The book moves to the issue of ontology in chapter 4, exploring the relationship between technology and one's way of being in the world. Heideggerian ideas about technology and notions of truth are elaborated and exemplified through everyday examples that reveal enframing for what it is. Sentiments of belonging are explored, towards an awareness of the differences between authentic and spurious understanding. Chapter 5 then moves forward from the idea of belonging to elaborate issues that arise with the globalization of culture through the movement of both goods and global relocations of people (and hence cultural practice). The connection of the global to the local is thus exemplified through vignettes that link global iconography and historical patterns to local experiences.

Post-global citizenship is the topic of chapter 6, which looks at the disciplinary aspects of acquired mobility. More explicitly, culturally specific disciplinary apparatuses are elaborated with examples that show their impact on subjectivity in the post-global context. Chapter 7 elaborates a theory of home for mobile citizens, linking mobile technology with ritualized and routinized performances of culture that can be taken throughout the world. The idea of ontological security is shown as a motivation for constancy in everyday performances of self.

Chapter 8 takes us back to the authenticity of being human, demonstrated in everyday moments of caring, concern, and human connection that transcend global/local specificities and technological inscriptions. The resilience of everyday people is situated within the context of technological performativity and advancement, as its inspiration and driving force, and at times situating technology as ancillary to the main drama of interpersonal interaction.

Italy and Chiapas, Mexico, are worlds apart. And yet, they have become inseparable from one another and are both important to the book in your hands. There is little topical correspondence between the indigenous agrarian lifestyle of one with the advanced capitalist bustle of the other. Within a very short time, however, the invisible connections of the global actor-network they share become revealed, through a conspicuous tree native to the mountains of southern Mexico growing happily in Boboli gardens, to a cross mounted on the parapet of a Catholic community church high in La Selva Lacandona. The Italian's nationalist appropriation of the tomato, an indigenous American fruit, is just one example of how the artifacts of global exchanges are made meaningful in localized settings. Technologies of religion are exchanged for technologies of food (who got the better deal is another topic entirely). As such, technology is no different than the "mein" (noodles) that migrated from China to meet the tomato in Rome, becoming not just a profound part of global history but an indispensable dietary staple in the present. Perhaps 500 years later, one might even be served spaghetti and tomato sauce in the mountain rain forest of Chiapas and think nothing of it.

Is it an accident that the largest contingent of supporters to respond to a wireless distress signal emanating from the indigenous peoples of La Selva Lacandona were Italians? Is it merely absurd coincidence? Is this global phenomenon all media effects, or is there something more profound informing these relationships? How does this global network form and reform itself? Observing the practice of culture in everyday life is one way to understand the nuances of sophisticated relationships stretched around the globe.

CHAPTER ONE

Global Technography: Ethnography in the Mobilized Field

The Importance of Everyday Life

It's a July Saturday in Seoul, South Korea. After shopping around together all afternoon for his new digital camera, my friend Sean suggests we go to Itaewon to get phone cards. Logically, the card with the best rate to the USA is available only in this area of Seoul heavily frequented by American soldiers. Telecommunications thus guides our action plan. A very short walk from the subway station, the mission is accomplished within minutes at a little nondescript convenience store. With that out of the way, we start walking down the main street of the area, passing rows of sidewalk vendors selling their imitation wares. We stroll by storefronts advertising leather goods, luggage, shoes, and even several outlet stores: fake Prada and Gucci alongside authentic Adidas and Nike. We finally arrive at a doorway leading up some stairs. "Let's go up here…I know this place…" Sean says.

We climb a flight of stairs to the second floor of Star Express and place our order at the counter to a blonde woman (probably Russian), but to me it looks like the older Korean woman behind her is the one who best understands what we're saying. She explains something to us and points in the direction of the tables, which we take to mean we can sit wherever we like. Thinking we're following her directions, we choose a table out on the balcony overlooking the street and sit down. After a while, I see the Korean woman put a tray with our order on the counter. Seeing me looking, she gestures with her hand to show it is ours. I walk over to collect the tray, but she stops me, explaining that we have to pay at the cash register in the other room first. I've embarrassingly run out of cash, so I go back over and tell Sean what we need to do.

It's now my turn to sit while he walks through the room and pays, he then brings our tray back. I need more sugar, so I get up and go back to the counter. Finally, we're somewhat settled. "That's a lot of running around for a couple of coffees..." I say offhandedly, thinking in a Certeauian mindset about how random people's pathways can appear even when they're guided by rather simple logistics. We talk a bit about various things, and as we relax into conversation my observational mind starts noticing things around me:

- At a table nearby are two non-Koreans, male and female
- Behind them are two Korean girls (I watch the girl talk on her Motorola Razor cell phone), later they are joined by one (who holds his phone in his hand for a while), then another man, they take up an extra table (Razor girl talks on her phone); then a few minutes later, another girl comes, then another girl, then they leave
- Behind me are two Korean women
- On the other side of the glass in the next room are three Korean women who have been shopping (they have their bags all around them)

Seemingly out of nowhere, I suddenly state, "All the Koreans have their cell phones sitting on the tables, and all the non-Korean's don't," including ourselves in the "non-Korean" category.

"Is that important?" Sean asks, "I don't understand why that matters."

"Is it?" I ask, not completely sure myself, but explain that it just stands out to my sociological eye.

We debate the issue of importance: What is important? What isn't? Who can tell? Why should anyone care? What difference does it make anyway? I notice a group of three Koreans walk by abreast a group of three Westerners. "Look down there...The three Westerners all have backpacks on, and none of the three Koreans have any bag on their shoulders...Is it significant? Is this just coincidence? And if it is coincidence, how do we even know that?" I ask.

"I don't know...if it is important, I don't see how," Sean says. I'm starting to get a little more excited now.

"This is how we describe life, man! This is how we illustrate a setting, create a context...You know, everyday life is boring and irrelevant! It's full of things that don't seem to matter! It is mundane...Something happens, and you notice certain details, and maybe it isn't significant, maybe it is, but what is important is how people feel about it..." In spite of myself, I'm in lecture mode now. "People don't like having their mundane shit disturbed, they will get angry and fight to keep their mundane lives intact! This is most of what culture, society, daily life, or whatever we want to call it is about, man! This is what makes things meaningful! It's not the cell phone that's important, it's the conversation it provokes that we're having right now that is interesting!"

"Hmmm...yeah..." he says, turning it over in his mind. "Yeah, I get what you're saying...So what do you feel like being back in Itaewon?" he asks, knowing intimately how I reacted to it last time I was there two years back. We seamlessly change the subject. I relate to him that I'm still impressed by the shabbiness of the area and those who habitate there, and I lament the fact that Itaewon's brothels and American soldiers will constitute the limit of experience and perception of Korea for many visitors.

"Yeah, but at least they got cheap phone cards!" he jokes, withdrawing his ringing cell phone from his jeans pocket.

Figure 1-1. A view of Itaewon from the balcony of Star Express. Amid the rummage of street vendors hawking fake goods in front of legitimate label outlets, meaning is created in the mundane performances of everyday life.

* * * * *

The path to this work began with the millennial keywords time, space, speed, and society, which together pointed to a technologically induced "speed crisis" that gripped critical theorists of technology during and after the Y2K scare. The threat of Y2K-induced global malfunction brought to the fore many serious issues that had remained unresolved in sociotechnical theory throughout the twentieth century, particularly issues with digital technology. This research addresses poignant questions raised in the discourse of both celebrants and dystopian theorists of technology:

> How can we really live if there is no more here and everything is now? (Virilio, 2000, p. 37)

> When precariousness becomes an endemic trait of the human condition and marks every facet of the place currently occupied in the network of social dependencies and commitments, one's hold on the present is most painfully missing—it falls in fact the first casualty. (Bauman, 2000a, p. 94)

How does the normalization of 3GHz processors in laptop computers or a reduction of the size of the cell phone compound the confusion of everyday technological experiences? The answers to these questions can be found in the examination of everyday technological experience—the "apprehension and use" (Carey, 1989) of technology by its end users. As a means of examining technological mobility from the user's experience of cultural and social change, this work responds to Peter Lunenfeld's "digital dialectic" that places "the human being at the center of technology" (2001, p. 45).

Potential Ignored

"Do you know about this?" Koji asks me, pointing to a strange-looking square about three centimeters on each edge. Koji has been guiding me around the city. He's in his late twenties, an engineer with an auto parts design company. Having spent the afternoon in Odaiba, we're now on the subway headed back towards Tokyo station where we will have some dinner.

"Hmmm...I've noticed those things on a lot of posters," I tell him. "They look kind of like a barcode, but with a different system...Is it a bar code?" I ask him.

"Not bar code...it works similar, but it's for information on the cell phone," he tells me.

"That's interesting...how does it work?" I ask. He takes out his phone, unfolds it, holds it up and takes a picture of the code box. He punches a few keys, then turns the phone so I can see it. The screen displays a map, showing the location of the business on the poster.

"It has other information too..." Koji explains, scrolling through some Kanji text, "But you can't read it, can you?" he asks.

I smile and shake my head to say "no" in response, then I remember that in Asia one generally agrees with the person, not the sentiment, and say, "Yes, you're right." With his phone now in hand, he scrolls through some pictures of his wife and kids like men might have shown off wallet photos in another era, telling me their ages and answering my query as to how long he's been married. "You have a wonderful family," I say. He then shows me some of the other features of his phone—video, Internet browsing, and instant messaging.

"Do you use these things very often?" I ask expectantly.

"Oh no!" he exclaims, "No one uses these things…in Japan, the cell phone is very expensive…"

* * * * *

The work presented in this book shows that examining technology as human-centered reveals the resilience of culture. Although our technologies change, we still make sense of the world according to our already taken-for-granted assumptions and experiences. The cell phone photo album has replaced wallet photos, but the ritualistic use of the images remains remarkably the same. We are often NOT mere subjects caught in a relentless current of merciless temporal and spatial reconfiguration. Rather, quite often, we knowingly and actively participate in provoking exactly the types of seismic shifts critical theorists such as Baudrillard (1988), Bauman (2000), and Virilio (2000) read as signals of doom. Since they themselves reveal all of their critical concepts—time, space, society, and speed—to be relational fictions, any actual perceivable "crisis" is duly ontological, within oneself and one's perception of one's own sense of being in the universe. Addressing exactly this issue in his seminal *Being and Time* (1996), Martin Heidegger's work provides a philosophical tool to situate people at the heart of technology (Heidegger, 1977; see appendix). In this approach to the contemporary situation of minitaturized, ubiquitously networked wireless communications, the subject matter of this manuscript is aptly described as a study of technological performativity in everyday life—a phenomenology centered on human/machine interactions. This is the typical domain of ethnographic research, but herein lies a methodological challenge.

Ethnography traditionally studies human interaction within a physically bounded "field" of study. This type of anthropology has been quite useful for studying "exotic" peoples in remote and urban spaces—in places that are physically locatable. However, the invention of cyberspace brought with it virtual reality, and with that came virtual culture, online communities, and cyber-society. These technologically enabled social experiences necessitated a reconceptualization of the ethnographic field. Numerous methodologists successfully dealt with this issue by treating cyberspace as a domain bound by the parameters of the technology, assuming a conceptually static plane of interaction (see Hine, 2000; Markham, 1998; Denzin, 1999). Chat rooms and other online social spaces were treated as spaces people entered and left, just as they would any ordinary building or town square. Many ethnographers took advantage of this approach, but then along came wireless technologies and the remediation (McLuhan, 1995) of formerly stationary technologies into portable devices. Wirelessly networked personal technology has changed the nature of the ethnographic field yet again, by deterritorializing technological experience. People now take the chat "room" with them everywhere they go, never really entering or exiting the environment. Ubiquitous mediation is the norm in the context of wireless connectivity. Wireless Bluetooth headsets and keyboards—peripheral devices—add an ever deeper confoundment to notions of wireless territoriality. How can one define a field when the users and technologies are constantly in motion and intermittent, and when the nature of the network itself has changed from the central-server model to distributed networking dependent on the active involvement of the users themselves? From previous ethnographic perspectives, the field has become unbordered and thus unmanageable. The answer lies in a reinvention of the concept of technography.

Breaking Frame in a Sacred Game

I look out disinterestedly past the bank of televisions mounted on the wall, through the large windows to the dark street. It's 9 p.m., Thursday night in the new "East Wing" of the IMPE building at the University of Illinois. I'm going through my routine workout and have arrived at the "cardio" portion of my exercise—the stationary cycle. My bored gaze travels back inside the building, up to the programs on the television screens: *Friends*...MTV... Baseball...What's this? An interesting development in the baseball broadcast, what I think of as America's sacred game. It's the Red Sox versus the Yankees. Some of the fans have riled up a Yankee outfielder during a play, appearing to have interfered with his outfielding duties when the Red Sox hit one to his area of the fence. The Yankee takes a swipe at one of the fans, and a melee ensues. Fans, players, and authorities are suddenly all on the field, fighting to reach the epicenter of the disruption. Various camera angles construct a multidimensional spectacle, and as the authorities begin to get the situation under control and the crowd to disperse, the director cuts to several replays that illustrate the fan interference that angered the player. After reviewing and discussing shots from a few angles forwards, backwards, in slow motion, and at regular speed, the director cuts back to the live scene. It's a bad choice of shot. The crowd has mostly dispersed by now, and the sparseness of the outfield crowd fills the screen with more than a few naked seats. Still, the camera lingers, dwelling on the fans who started the ruckus. As the camera zooms in on them—a group of three men in particular—I notice they're all holding cell phones in their hands. They look like they're examining pictures they've snapped, possibly sending them to friends, or even each other. They're individually off in their own separate worlds, oblivious to the surveillance of the TV network's cameras, focused intently on their own business, yet seemingly unified by their cell phone performance. Altogether, it makes for very uninteresting television, but around them I notice several other people striking similar poses with their phones before the camera cuts away. "The enframing revealed for what it is…" I reflect, deciding this is exactly the kind of timelessly uncomfortable moment Heidegger was writing about. Thinking of Erving Goffman's Frame Analysis as an appropriate way of understanding this eruption of truth, I break frame myself at this point and look around to see if anyone else witnessed this strange moment in sports television. "Goffman…yeah…of course, I have to think of a Canadian..." I smile silently to myself.

* * * * *

As long as technology continues to be treated as an intruder into our everyday lives rather than being recognized as an inseparable part of it, our understanding of the world unfolding before us will remain ignorantly, perhaps sometimes even dangerously, flawed. Many anthropologists and social scientists concerned with recent sociotechnological developments have been studying "the arts and crafts of tribes and peoples"[1] by drawing from classic Goffman texts (see Goffman, 1959, 1963, 1971, 1974). The tradition-

1 As defined in <http://www.americansubstandard.com/definition.php?word=Technography>, <http://www.drwords.com/define/Technography>, < http://www.thefreedictionary.com/Technography>.

al symbolic interactionist approach is perhaps a natural and obvious way to proceed in dealing with the confoundment provoked by the breaking of public/private social frames that comes with the ubiquity of portable wireless technology. Numerous works have used this perspective to explain particular human-centric performative dimensions of technological behavior that become more noticeable with the normalization of virtual experience and the reinvention of electronic space as mobile territory (e.g., see Dell and Marinova, 2002; Ling, 2004; Humphreys, 2005; Waskul, 2005; Gotved, 2006; Katz and Sugiyama, 2006; Soukup, 2006; Robinson, 2007; Couch and Liamputtong, 2008; Schick, 2008; Whitty, 2008). Important as they have been in helping us understand these new spaces and behaviors, technology continues to be treated as a prop or dead "thing" somehow situated outside the arena of social interaction in Goffman-informed research, even as it works as a shepherd to contain and manage actors in a field of study. This approach has taught us a lot and, there remains much to be explored from this type of work. However, it can tell us only part of the story of technological experience, leaving out important new aspects of culture and society brought on by the unmistakable vitalization of what was once considered inanimate. Reawakening of the term technography and retheorization of the ethnographic field to contend with ubiquitous electronic space allows ethnographic documentation of our very personal, intimate experiences with technology, and the myriad relationships, thoughts, and feelings technology participates in. Global technography helps us to show that technology is a dynamic and dramatic participant in the performance of everyday life. It shows us what critical geographers such as Lefebvre (1991) and Soja (1989, 1999) have known for some time already: that the props, backdrops, and containers of our interactions practice everyday life right along with us. Technology spatializes practice wherever it is found to be working, just as Carey (1989) described the human demarcation of space by movement through it. The field thus temporarily "erupts" around the device every time it labors, and attention to the ever-changing vicinity of the device itself thus becomes the defining parameter of the field of study.

Cell Phone Lost and Found

It's Wednesday afternoon, about 2 p.m. I'm walking north past Neuman Hall, a Christian residence on the University of Illinois campus. It's sunny, but still cold, like spring days often are. I'm on my way to the wireless-enabled Illini Union building to do some writing, my laptop in my backpack.

I randomly look to my right, and to my surprise spot a cell phone on the ground just off the sidewalk under the hedge. "Hmmm..." I say to myself, reaching down to pick it up. As I resume walking, I look at its worn leather cover. It has a belt clip on it. "I knew those things couldn't work very well..." I tell myself. I open the clamshell looking for clues. It's a Sanyo phone, with a camera, and apparently a Verizon phone plan. I decide to explore the contents of the phone. The screen displays a picture of a young white couple; to me, they look like the epitome of "normal." Almost automatically, I find myself looking through the camera's stored pictures. There's nothing strange, just typical goofing around snapshots, and a few more seriously posed shots. I look in the address book, scrolling through numerous entries. I land on one named "mom." "She'll know what to do..." I say, pressing the dial button.

"Hello?" I'm greeted by what I'm predisposed to think is a momlike voice.

"Hi…Um…you don't know me, and I don't know you, but I just found this phone on the ground, and I found this entry called mom, so I thought I would see if I can call you to get this phone to whoever it belongs…"

"Oh…Oh, I see…OH!…Thank you so much for calling! Oh, my daughter will be so happy you found her phone!" she responds.

I ask, "Well, can you get in touch with your daughter, and let her know I have it?"

"AAAAmmmm..." she says, thinking. "Well, she'll call me, I'm sure of that, so maybe you can hold onto it and when she calls I'll tell her to call the phone, and you can make an arrangement for her to pick it up from you…"

"OK, that sounds reasonable," I reply, "I'll wait until I hear from her then." We say goodbye and hang up. I continue on my way to the Union. "I don't have time…I don't really want to have to meet this person…And answer her phone..." I think to myself. "Who knows when it's going to happen?" I decide on a new course of action. I hit the dial button, and "Mom" is quickly back on the phone.

"Listen," I say, "I'm going to leave this phone at the front desk of the Illini Union…Your daughter can pick it up there…"

"Oh, I see…hmmm..." she says suspiciously.

"Don't worry, I'll tell them the owner knows it's there and will come pick it up," I reassure her. "I just think with my schedule this will be easier than trying for her and I to meet up personally..." I tell her.

"OK then, thank you for helping..." she says.

"No problem..."

"Bye."

"Bye."

As promised, I take the phone to the front desk of the Illini Union and explain the situation to the desk clerk. They agree to my plan. My good deed done for the day, I proceed to the Courtyard coffee shop area, unpack my laptop, and resume where I left off writing about technology and everyday life.

* * * * *

There has been a precedent in the quest to study technology as participant in everyday life. Actor-Network Theory (ANT) has been a reliable method of conducting ethnographic analysis of technology for more than two decades. Technography as a category of inquiry historically precedes ANT, appearing in anthropological literature in the nineteenth century (see Kien, 2008), and my use of the term here does not challenge or seek to displace ANT. Rather, I seek to embrace and build on what ANT has taught us: that the relationships between humans and machines are complex and messy but are ultimately possible to document and quite enlightening in what they reveal. My motivation to contend with the experiences of mobility and globalized ubiquity of technology and the ANT goal of collapsing modernist inside/outside dualisms share much in terms of theoretical goals. However, once that dualism is effectively collapsed, ANT is pushed into unfamiliar territory; one might say, ANT is forced to have territory, contrary to its typical emphasis on "things" and the abstract arrays that produce them. Attention to the network array ironically suggests keeping a somewhat macro-perspective sociological imagination even while acknowledging the ensuing paradox. Latour (1986) once critiqued sociologists as obsessed with the glue that holds society together, suggesting society—the rightful subject for social science—is that which the glue holds together. We can now see that it is neither just one nor the other, but both and perhaps much more that constitutes the social. The choice of topic is arbitrary, perhaps predetermined by the individual concerns of the researcher. Hence, my own concerns, informed by the circumstances of my own sociotechno experiences, are with human-machine relationships themselves from an intimately experienced, globally mobile perspective. My methodological aspirations are firstly to offer sociotechno researchers an example of how we can embrace advances in qualitative methodologies and move past the traditional interactionist approach, and secondly to show qualitative researchers in general that acknowledging technology as a participant in everyday life does not have to entail the sensationalism of science fiction.

To sum up, four intellectual strands come into harmony with each other to make this type of work possible: Actor-Network Theory, Heidegger's philosophy of technology, Norman Denzin's approach to Interpretive Ethnography, and my own participant-observer position as a global citizen—an "inside outsider" as it were. As the dominant agent of this document, I forge an alliance between Latour's (1988a, 1988c) premise that machines are built to enact social programs, and Heidegger's (1977) aforementioned assertion that humans are the "engines" of technology, inseparable from it in our profound reliance upon it. This coaxes an enthusiastic embrace of the ANT, which bridges philosophy with ethnography, connecting Heidegger's philosophical position to everyday performativity. But for the reasons previously mentioned, this requires advancing qualitative approaches to studying and reporting technological phenomena, since in many respects, qualitative technological researchers haven't kept pace with evolutions in qualitative inquiry as a methodological field. Finally, acknowledging myself as a global citizen with a slightly higher than average technological savvy reveals that I have been uniquely positioned to synthesize these various strands and innovate a technography that can show some effects of the technologically enabled mobility that characterizes life in the context of globalization. So then…how do we actually do it?

Flamenco, Cell Phones, and Puppy Love

It's a Wednesday night, mid fall semester. I was lucky to get tickets just two days before a sold-out flamenco performance at the Krannert Center for the Performing Arts in Urbana, Illinois. I'm familiar enough with flamenco to know this show has its moments of brilliance. It is a Spanish company on tour in North America, and for the most part, the audience seems to share my appreciation of their talents. That is, until the second half.

The only available tickets when I called were near the back of the famous "almost acoustically perfect" hall. The spectacular first half has concluded, and intermission is almost over. As people stream back into their seats, I note that the row directly in front of me is filled mainly with high-school students. "Must be a Spanish class trip or something..." I speculate.

I've seen this very type of outing at a flamenco bar back in Toronto; some Spanish language teacher decides to take their students on a field trip to "experience" some Spanish "culture." I recall my analysis of that much more cynical time in my life: "Of course, what is actually happening is they are merely witnessing the appropriation and reenactment of Gitano folklore in a rather contingently relocated context, to be consumed along with a menu of other items exotically tailored to best convince people to part with their cash in hopes of owning part of the 'magic' they have witnessed ..." Be that as it may, in fact, I am about to be treated to a much different type of performance than that happening on the stage.

Once the curtain is drawn and the second half begins, it becomes painfully obvious that two acts is one too much for the school kids forced to be seated in front of us. Where they were previously very well behaved and attentive, after only a few minutes their cell phones have suddenly appeared in their hands. And not just a few, but almost the whole row. For those of us behind them, the pale glow from their screens easily distracts us from what is happening on stage. Some put their phones away again, but the three in front of me seem to have something else going on. The tall athletic-type guy in the middle is flanked by a blond girl on one side and a brunette on the other. They whisper to each other, the guy acting as the conduit between the two girls. I notice how he plays them, first looking at one's cell phone, then the other's, reading the clever text messages coming and going. They are all instant messaging their friends, and I easily focus in on their text over their shoulders. They are reporting where they are, and what they're doing, and how BORED they are already. The text messaging play continues, but another dimension is quickly added in. Due to boredom, the girls occasionally find it necessary to rest their heads on the guy's shoulders. I can't tell if each knows that the girl on the other side of him is doing the same, but I do notice that he is good at directing their attention to either his or their own screen whenever they're in danger of looking past him to the girl on the other side. Thinking of the three typical levels of media analysis, I begin to wonder silently, "Is this audience/text, audience/medium, or social relations around the medium ... ?" I decide it involves all three. After a while, I get bored with their performance myself and turn my full attention back to the stage. "A little taste of teen culture...," I think, pedantically pondering the relation between the evening's performances, one on stage, one in the seats.

* * * * *

Investigating humans as the engines of technology requires an alertness for those subtle moments when truth erupts, in which the enframing of everyday life is revealed for what it is. It may happen in places one might expect, but more likely it will happen in the course of mundane, everyday activity such as working out or consuming another type of media. Understanding technology and humans as actants and actors in "messy networks" requires working through the contradictory goals of the necessity of structuration, while at the same time accepting that the arbitrariness of any closure that imposes order is of necessity inaccurate. Autoethnography (Ellis and Bochner, 2000) as a reporting method allows an incredible depth of detail to show how everyday life is experienced but at the same time begs a constant self-evaluation in which the author must persistently question how much revelation of one's own life and that of others' is acceptable, not just now, but in the future. Ethnography does have long-term effects on relationships and organizations (Simmons, 2007).

The Actor-Network Theory method of "following the actants" and elaborating their actor-network through their travels also presents challenges: it's not always easy to "go where they go," and it isn't always a simple task to recognize some of the subtle and tenuous alliances they enlist and are enlisted in. However, following the wireless laptop and cell phone in "glocalized" settings produces one unmistakable finding: the root physical technology for both wireless devices is the global telecom network that comprises the most stable material aspect of the network,[2] while wireless devices themselves act as conduits/portals for individuals to ally with the network in various guises. In this particular sense, although the content has evolved and the potential for connectivity has exponentially increased, the present incarnations of wireless mobility represent an evolution (albeit a rather massive and dramatic one!) of the global telephony system. Nonetheless, although insinuated into incredibly nuanced performances of everyday life, the ways we qualitatively ally with the physical network often continue to determine which of its potentialities will be realized.

---

2 For an ANT definition of "relational materiality," see Glossary, p. 175.

Wireless Scavenger

Monsoon season in Seoul. I write an email to my brother about the difficulties I've been facing in trying to dry my clothes without a clothes dryer in the constant rain and high humidity:

> Here I am online! I found that my computer can use one of my neighbor's wireless connections, so I don't have to worry about getting an Internet account. I feel more connected now. Yep, here I am in my little top floor apartment. It's 8 am, I'm sitting here having my morning coffee and reading email, just like I do everywhere in the world. In some ways it's a lot like being in Toronto, in other ways it's obviously a different country. They drive motorcycles on the sidewalks here!
>
> Monsoon season is upon us. Nobody uses clothes dryers here, so the building only has washers and then people use clothes lines on their balconies to dry the clothes. I washed my clothes on Tuesday and put them on the line, then went to work. It started raining while I was at work, then rained all day Wednesday. I left my clothes out, thinking it would have to stop at some point and they would dry. It was still raining Thursday afternoon when I came home for my afternoon break, so finally I decided to move everything inside. I had socks, underwear and shirts hanging everywhere. It was so humid in here, they still hadn't dried by Friday afternoon, but it had stopped raining for a while, so I moved everything back outside. By the night time (last night), I looked out just before going to bed, and sure as hell, it had started to drizzle again. I ran out to collect my stuff, and caught myself muttering under my breath, "Fuck... I'm never gonna get these fuckin' clothes dry..." So here I am Saturday morning, sitting at my little dining table typing on my little laptop, surrounded by my damp clothes! It's stopped raining again, but it's still overcast. Do I take a chance, or just wait it out inside? I'm not sure, but the most surprising thing for me is to learn that I sometimes swear and mutter under my breath ;)

This particular morning I can't get an Internet signal inside my apartment, but have found it possible to pirate one or two connections from the back edge of my rooftop balcony. I use it to check and send email that I compose inside. When it rains, I go out with my laptop balanced on one arm under my umbrella, walk to the edge of the building, and stand there like that until I've had enough Internet time for the files to transfer. Sometimes I even have a short one-handed IM chat. I must look totally insane to everyone! The insane Westerner out with his laptop in the rain... I download all the news stories I want to read in separate browser windows, then go back inside and read them. I maintain my morning ritual without interruption, even on a different continent.

* * * * *

A fifth methodological strand must also be mentioned here: the vignette reporting style, which is one of the most effective pedagogical tools for helping situate a reader into an empathetic reading of a text (Tettegah and Kien, 2003a, 2003b). Use of vignettes allows the researcher to textually portray social phenomena in a way that approximates the experience of it, rather than adjudicating it from the point of view of an outside, expert observer. This sits well with Denzin's (1999) notion of "authentic understanding," which invites the reader to textually experience researched phenomenon, feel its impact, and thus authentically understand its importance rather than just be told about it. I have elsewhere described the resulting textual style as "autoethnographical fiction" (2007), taking a cue from Ulmer's (1989) method of "Mystory" that simulates the aesthetic styling of video-editing practices to construct a textual portrayal of a chosen phenomenon. The overarching rationale of my stylistics takes to heart the ANT directive: to produce a text that "provides" time rather than steals it, in terminology that the actors themselves would recognize, use, and understand (Latour, 1988b).[3]

Pirates of Hou Hai Lake

I'm with a Chinese friend at the far end of Hou Hai Lake in Beijing. We're seated outside on luxurious green sofas and are served by the staff from the bar across the lane. He has recently returned from the United States for his vacation. His wife has stayed behind to take care of some business but will arrive within a few days. He pulls out one of the products of his day's consumerism: a new cell phone for his wife. "It has two SIM cards ..." he explains, "so I can get her an account in Shanghai as well as Beijing, and she won't have to carry two phones..."

"A lot of people carry two phones here, eh?" I ask.

"Yes," he replies, "I'm actually carrying THREE now...It's crazy...One Beijing digital, one Shanghai digital, and my Wei Wei" (Hello? Hello?; a nickname for a cheap analog cell phone that frequently drops its signal, whose number many Chinese people would give out for incoming calls).

3 For further explanation of ANT methodology, see Glossary, p. 169.

As he speaks, one of his phones alerts him that he has a message. "Just a minute...I have to reply to this message ..." he excuses himself from the conversation for a moment. I take the opportunity to look out across the placid lake to the opposite shoreline, ingesting the night scenery of countless red lanterns assuring people that food and drink are plentiful. From the middle of the lake, I hear a girl's voice cheerily shout something indistinguishable, but I can tell it's in English. I scan the multitude of paddleboats and then direct my attention back to my friend, who has just snapped his phone shut.

We relax and talk for a while on the couches, our space occasionally interrupted by his various phone calls. At one point, he stands up, excuses himself, and then strides across the alley into the bar to search out the facilities. I again look out across the lake. I realize I'm absentmindedly listening to the English-speaking woman's voice again: "Dahling...we must get more to drink..." followed by a few words of Chinese that I can't understand. I realize the origin is a paddleboat that seems to be heading straight towards me. I watch the boat slowly but surely emerge from the darkness into my range of vision. I make out the figures of two young women and a man. The two women maintain a constant chatter between themselves in Chinese peppered with what sounds to me like faux British phrases ("Dahling..." "Smashing..." "Adorable..."). As they get closer, I make out a faint glow on their faces and realize that the man has a laptop sitting in front of him. They have my full attention now. I reach into my backpack and pull out my camera, zooming as far as I can to capture this surreal moment: "Laptop on the Lake," I mentally name the photo. They paddle the boat closer and closer, until they are almost right below me. I hear the soft strains of music coming from the computer. The two women climb out of the boat and crawl onto the shore right behind me, cawing away to the service people about the beer they want to buy. They are well dressed and friendly, but my attention is on the laptop. The man, in a suit, waits in the boat and attends to his computer. The current causes the boat to drift slightly, and it turns so I can make out what he's up to. I can see he's using Windows Media Player as a jukebox. He clicks another window to the front, and I see the telltale progress bar: he is downloading music, even while out on the lake. He then clicks to an email client briefly, then back to the media player, where he inserts the next song in the cue. I zoom my camera in on his computer, using it as a telescope of sorts. On my camera's monitor, I note that I can actually make out the little antenna of his PCMCIA card. "Must take forever to download music with that…Maybe it's one of the new high-speed ones..." I reflect, but then, I had myself downloaded some files the same way the previous night. Suddenly, my friend is back, sitting across from me again.

"You see that?" I ask.

"What's that…a notebook?" he asks, looking out at the paddleboat. "That should be good for your research..." he comments.

"Yeah, I think it is…He's downloading music…I got a few pictures… I'm not sure what this is about, but…I guess I'll figure it out later..." I give him the "doing research" smile.

"Downloading music? Really? That's kind of strange..." my friend comments.

"Yeah, I guess he's one of these businessmen with the PCM card, you know, to use the cellular network with his laptop…Are those girls from Hong Kong or Taiwan or something?" I ask him.

"I don't think so…They have Beijing accents, but they keep saying these strange English phrases..." he observes.

"Yeah, that's why I thought maybe they were from Hong Kong, you know, because of the British influence..." I say. "Maybe they spent time in Australia, or New Zealand?" I speculate.

Beside us, they are reboarding the paddleboat with a newly acquired six-pack of Beck's beer even as we speak. They don't seem to notice we're talking about them. We turn our attention to other matters, but I talk with one eye on the paddleboat as it slowly creeps out of my range of vision. Boat now out of sight, I hear one last faint call from the middle of the lake: "DAH-ling…"

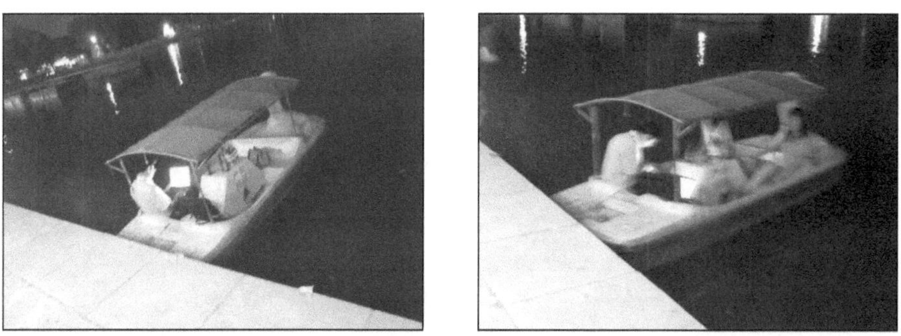

Figure 1-2. Laptop Pirates on Hou Hai Lake. Left: A man downloads music in a paddleboat while waiting for his companions to return with beer. Right: The party paddles away, speaking a mix of Beijing Chinese peppered with strange quasi-British phrases.

CHAPTER TWO

Network and Power: The Global Landscape

The frantic abolition of all distances brings no nearness...Short distance is not in itself nearness. Nor is great distance remoteness. What is this uniformity in which everything is without distance?
—MARTIN HEIDEGGER

Being in the Global Network

It's a sunny midday spring to the university. I'm riding my hybrid bicycle (made in Quebec, purchased in Toronto) south on First Street in Champaign, Illinois. I carry my laptop (made in China) in my backpack, and my cell phone (made in Korea) in my jeans pocket. I navigate the usual smattering of indifferent domestic and imported automobiles commuting through Champaign-Urbana's streets. Flying past strategically located Mexican, Indian, and Italian restaurants, without warning I find myself suddenly experiencing a feeling of incompleteness, even as I resort to the tactic of using the crosswalk to negotiate the intersection at First Street and Springfield, past an Asian grocery store. In spite of the close physical proximity of my wireless network appliances, I can't help feeling that I want to be connected right now, right here, this very moment. This temporary disconnection from the global communications network is an interruption in my preferred state of being. My ontology has become deconstructed. Have I become dependent on my appliances for my sense of security, enslaved by the technologies I use? Is it a technologically induced coma that I desire—McLuhan's fatalistic "narcisis narcosis"? Is it perhaps the joy of dwelling in my global social network that I crave—McLuhan's utopic "global village"? Has technology, in fact, propelled me to a new frontier of interactive personal freedom? I want to know if I have any new email right now... I want to know if there is someone to chat with on Instant Messenger this moment...I want to know if there is an online bargain (i.e., a pricing error) being reported on a web site that I'm missing out on right at this instant...I want to know right now if there is some horrific disaster unfolding somewhere, affecting the world I and those who I love inhabit...I want to be assured that I am cared about, and I want others to know that I care about them, right now, wherever they are.

Perhaps in this moment I am for some undiscovered reason overwhelmed with sentimentality and simply seeking a feeling of reassurance through mediation. Regardless of how temporary, disconnection from the global network is implicated in my feeling of loss—specifically, a loss of connective potentiality. No less important, my own technological expectations are also part of the matrix of my ontological experience. After all, I should be able to be connected right now! Isn't that part of the information age promise? I find myself imagining a wearable interface such as I've seen Steve Mann sporting. I imagine myself cycling along in the sunlight with retinal projection readouts in my peripheral vision, and a wearcam streaming an upload of the road in front of me. I imagine myself in conversation with my international world while I cycle through Campustown. I want to be the picture of the technologically perfect mobile subject. More so, I want to be the technologically perfect mobile actor.

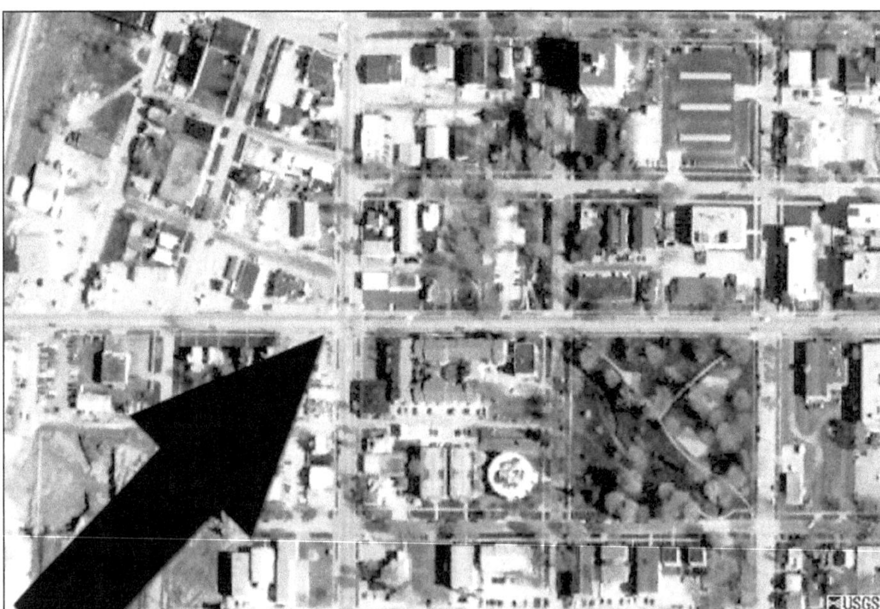

Figure 2-1. First Street and Springfield Avenue, Champaign, Illinois. Is a feeling of disconnection while in transit symptomatic of technological addiction, or longing for global village? (From terraserver.microsoft.com)

* * * * *

Even though technological equipment terminates remoteness, it does not actually gather things closer or move one nearer. The initial distance between things is implanted in the media themselves. For example, television brings remote things within view, but at the same time it conceptually holds them away and keeps them from being closer. Whether equipment or art, a thing is something that does what it is meant to do, presencing by gathering what it gathers to do its work. However, with modern technology, instead of bringing far things near to be authentically experienced, everything becomes imbued with "uniform distancelessness" (Heidegger, 2001, p. 164). Conversely, there is no longer a distinction possible between near and far; the words become meaningless, obliviating the requirement of distance needed for conceptual isolation and abstraction.

The context of this study is the confluence of three distinguishable phenomena: the network, the media, and the everyday performances of culture that bring these into play with each other and with human actors. Globalization is often described as constituted to a significant extent by networks, places, and mobility. Castells (2000) goes as far as to state that network is what makes globalization possible, while Hardt and Negri (2000) theorize politics of the twenty-first century to be a globally networked circumstance even in the minutia of interpersonal relationships. Within this global network array, "place" is used to name localized sites of activity and agency. "Network" is thus a spatialization of relations between places, giving rise to the conditions that make globalization possible. The third component, mobility, portrays movement between places of people and/or information and other goods within networks. Altogether, a picture emerges in keeping with Doreen Massey's description of "power geometry" (1993), in which places, networks, and movement produce geometric spatializations reflected in geographic arrangements of power.

Seeming to build on the historical precedent of McLuhan's (1995) "Global Village" concept, Baudrillard appears to have been one of the first to recognize the importance and sophistication of the then emergent concept of global technological networking. The *Ecstasy of Communication* (Baudrillard, 1988) evokes an extreme McLuhanistic theorization of networked appendages. While Baudrillard's attention in this work is clearly on network, the concept of place is hinted at as a seemingly inert human hub of networked activity. He describes this technological embodiment as "private telematics" in which "each individual sees himself promoted to the controls of a hypothetical machine, isolated in a position of perfect sovereignty, at an infinite distance from his original universe" (Baudrillard, 1988, p. 15). Whether intentional or not, Baudrillard's use of the term "distance" hints at a Heideggerian-informed concern with modern technology in which "de-distancing" enframes the truth in such a way as to keep it hidden and unknown (Heiddeger, 1977)[1]. Baudrillard theorizes the end of the individual in this context: the self becomes a postmodern grotesque, a mutated human/technology network assembly not unlike Callon and Law's (1995) notion of the hybrid collectif: the sum total of technological appliances that constitute it.[2] In keeping with his theory of simulacra, "the real" becomes something apart from the self, something so far removed from its origin as to be no longer recognizable as such. Dealing more directly with network as a technology, Baudrillard explains the practical danger of such networked ontology; that the formation of a single global network makes the entire system vulnerable at any single point (Baudrillard, 2002). Commenting on the 9/11 Al Queda attack on the United States, he explains that such terrorism is a "situational transfer" in the global network, disrupting the multiplicity of generalized symbolic ordering by imposing an "irreducible singularity" on the system (Baudrillard, 2002, p. 9). This use of actor-network terminology (see Law, 1992, 2002)[3] highlights the aesthetic struggles over symbolic ordering that can be taken as indicating the places where power is enacted in the global network. This aesthetic struggle and focus on network is also a concern of Baudrillard's colleague Paul Virilio.

The Story of Two Davids (Part 1, Scene 1)

In the context of network power, hegemony can shift so quickly. Though it's two years since my last visit to Seoul when I worked at this same English language prep school for a summer job while doing preliminary research for my dissertation, the general appearance of the area looks familiar, However, some changes are easy to spot. There are three new Chinese language schools in the vicinity of Hoyah Academy, and school enrollment in general is significantly down. Still, I take a teaching job for five weeks just to pick up some extra money, and I quickly settle into the already familiar routine of

1 For more on Heiddeger's philosophy of technology see the appendix in this book.

2 For an ANT definition of "collectif", see Glossary, p. 170.

3 For an ANT definition of "singularity", see Glossary, p. 175.

helping Korean middle-school students get ready to study in North America.

Most of the staff has changed over, but in addition to Johnny (who started as an assistant and was promoted to HR manager since my last term of employment), I reunite with Se Ho and Robin, two Korean teachers who have been working there since the beginning of the school several years ago. Unfortunately, they look horrible—trapped, and with no prospects of change. Like Robin, Se Ho also has an English name (David), but he had told me previously that I should call him Se Ho, as it is more friendly. I don't smoke, but to be sociable I join Se Ho on the rooftop for a cigarette break. While I watch him smoke he blatantly tells me he's stuck. He knows his problem but doesn't seem to have a solution for it. He asks what I've been doing, and I tell him very straightforwardly that I just came from Tokyo and will go on to Beijing. I can see him choke back the bitterness, fake a smile, fight with himself to accept for the moment where he still is. I feel guilty for telling the truth. He tries to put on a philosophical face, and we return inside.

One day, on a whim, I take a break from my other lunchtime wireless hotspot, the Subway sandwich shop. While standing at the counter, I decide I would rather not spend this lunchtime in that environment, so I get my order to go ("take out" as they say in the local anglicized idiom). I trudge up six flights of stairs to the top floor of the school and pass through the staff-room door. I say hello to the room in general, take a seat at a desk, take out my computer and power it up, and start to eat while checking my email. There are about half a dozen other teachers in the room. Some are Kyopo (Korean Americans) and some are native Koreans, but I'm the only "authentic" Westerner in the room at that moment. Se Ho is sitting at a desktop computer with American easy-listening music streaming out of its speakers. He occasionally sings along a line or two. In feigned friendliness, he states pensively, "Aahhh...Grant...Well...my friend...I see you are having your lunch here today...," I tell him it's just a sandwich and then take a disinterested bite and keep on chewing silently. I hate the music he makes the rest of us listen to. He asks about my sandwich, not to ask if it tastes good, but more as a comment: "You like eating this?"

"It's just a sandwich...," I tell him with a touch of annoyance, trying to give him a signal to leave me alone. I can feel his animosity leaking out of his words and disdainful stare. I know I am a challenge to his status. He is a little bit older than me and has been with the school many years but doesn't have nearly the credentials. His English is fluent, but he is still far from a native speaker. Having never traveled outside of South Korea, he perhaps feels most threatened when he presents himself as a worldly person, knowledgeable about sophisticated cultural nuances of the West. I know that he is challenging me to somehow bolster his rank in the teaching pool.

"Yes...there are a lot of places to buy sandwiches around here...," he tells me.

I deploy a diversionary tactic: "What are you listening to?" He describes briefly the music he's subjecting us to, as if I don't know anything about it. I refrain from mentioning that before entering academics I spent ten years working in the music industry. I just nod and keep eating, then at the first opportunity, I excuse myself from the conversation by stating that I must work on some test preparation on my laptop. He stands, proclaims that he will have a cigarette, and exits the room. He comes back a while later, but busies himself with his own prep work before sliding down to teach his class.

* * * * *

In newly globalizing territories, globalized "Western" icons may be taken as symbols of modernist anxiety, a contradiction that serves to mark both progress and integration into the global system, at the same time inducing a sense of loss, confusion, and unequal power in the same global system (Xia, 2008). It is perhaps fitting that, being an architect, Virilio was one of the theoretical pioneers concerned with mobility. A book quite possibly ahead of its time, *Speed and Politics* (1986) explicated the relationship of mobility and power in the context of physical space, explaining that the amount of control over movement of people and goods through space is tantamount to the power that political formations can rightfully claim to have. In this perspective, Virilio locates power in the physical structures and conduits (places) through which people must pass, and in the structured social apparatuses (police, army, etc., aka "machinations" in Actor-Network language) that respond to "abnormalities" that might occur in those places. Virilio was prolific in exploring the nuances and impacts of his theory through the past two decades, but his writing in *The Information Bomb* (2000) nicely brings into focus his understanding of the contemporary situation, relating networked ontology similar to Baudrillard's description with not just physicality, but also temporality. Virilio acknowledges different temporal experiences enabled by technology. In another seeming reference to Heidegger's philosophy of technology, Virilio, like Baudrillard, explains presence as relational to and dependent upon distance. He uses the term "tele-surveillance" to describe the ability to observe a McLuhanesque global village (Virilio, 2000, p. 13). Screen media acts as a "domestic telescope," allowing the experience of viewing the future and past as well as the present (Virilio, 2000, p. 16). Such "tele-vision" exposes and invades the domestic space of individuals, giving rise to panoptic global vision. Virilio calls the present an "*axis of symmetry* of passing time," the geometrical origin of which is an "omnipresent centre" that controls "the totality of the life of the 'advanced' societies" (Virilio, 2000, p. 126). Being embodied thus means being a time/space nexus, a place of juncture in one or more networks much like Baudrillard describes, but with the addition of the virtual time machine as part of the network apparatus. This seemingly explains how we can have completion of what Law (1992) calls "network consolidation" beyond what Baudrillard described.[4] Such ontology emphasizes aesthetic ideals, stressing qualitative experience rather than quantified size and/or geography as a mathematical project (although both are used as tools towards creation of experience).

Virilio describes global warfare in this networked situation as being waged by a eugenic program of prevention rather than killing, the preemption of life itself for that which is undesirable. Power is in the ability to participate in birthing aesthetic order. As he later describes, global war is a battle over the world's images: The "optically correct" succeeds the "politically correct" (Virilio, 2002, p. 31). Democracy becomes the circulation of aesthetic representations through networked tele-visuals and tele-presence. Here it is once again control of movement—the mobility of signifiers to be exact—that becomes the indicator of power. Latour (1988b) uses the word "potency" to describe such control of the aesthetic order.[5] This emphasis on signification is also discussed by Manuel Castell—one of the fundamental authors contributing to the globalization canon.

The Story of Two Davids (Part 1, Scene 2)

I'm rushing down the hot, crowded sidewalk, about to commit to the ultimate fast food performance and get takeout at McDonald's. It's now Thursday, and I've spent too much time filing my Chinese tourist visa application in preparation for the next stop in my research schedule. I still have a pile of grading and my class prep to finish before I start teaching at 4 p.m. The

4 For an ANT definition of "network consolidation", see Glossary, p. 173.

5 For an ANT definition of "potency", see Glossary, p. 174.

McDonald's outlet is very conveniently right around the corner from the school, but my plan is cast into doubt when I get inside and see the crowd of people standing in line. It looks like a 10–15-minute wait. The air-conditioning is great, but I can't justify it. Luckily, through the side window I spot a gyros stand out in the alley. I make my way to the stand and place my order for a gyros and a Sprite. After a minute, two Middle Eastern-looking men serve it up alongside their other customer's order. I think momentarily of one of my "foreign" friends in Korea, Tamer, who is an Egyptian. For the first time, I realize there might be more people from the Middle East in South Korea than just him. Muslims seem pretty invisible, and maybe with good reason, since South Korea has the third-largest military force in Iraq. I pay the bill with a *komapsumnida* (thank you). Food in hand, I make my way hurriedly up the street and into my workplace. I make the loathsome ascent to the top of the building and take my traditional lunch seat in the staffroom.

As usual, I take out my laptop, putting myself in motion towards what is now a somewhat routine afternoon pattern. And, as usual, Se Ho is at his post, streaming easy-listening tunes to the silent derision of the rest of us. Some people have escaped into the sonic landscape of their own headphones, and I contemplate following their lead but decide I should first finish eating. On cue, Se Ho turns with his smile of scorn and sneeringly states, "So...Grant, my friend, you are having your lunch here again... Another SANDWICH I see...," he emphasizes the word, expressing his disapproval. "Looks like good enough food for a foreigner...," he comments with contempt. His emotional scar is laid bare in his bitterness and sarcasm. The poor, hated sandwich—symbol of the Western culinary invasion.

Of course, I hadn't given it any thought beyond the amount of time I had to budget that afternoon, but I try to be polite and answer anyway: "Actually, this is a *gyros* sandwich...it's Mediterranean, made with pita and lamb. It's tasty...I got it around the corner if you want to try one..."

Se Ho gets immediately defensive: "Oh...oh no...I prefer to eat Korean food... You know, we are in Korea...most people here eat Korean food..."

I actually laugh out loud at his nationalist indignation, doubly so to mock his decision to make me his target. With a big smile, I calmly, pensively say, "Oh, so, the Subway up the street, and the McDonald's around the corner, and the Starbucks down the block, and the Baskin Robins, and the KFCs, and the Burger Kings, and ALL of these Western stores are ONLY supported by the Westerners in Korea? Wow…There must be a lot more of us here than I thought…," I take a bite and chew. By "us" I mean Westerners. What it must be like for the men who actually served me around the corner I can't imagine. He stammers a weak explanation that some things are changing, lamely blaming "the kids," but I can't stop. I know that it is breach of Korean manners to comment on someone else's choices, just as it is in Canada. I also know it's a "lose-lose" situation, that neither of us can come out of this feeling good, but I've grown weary of him making me defend my choice of lunch in front of everyone, and I won't suffer it any more. I continue, "So, you think this food doesn't *belong* here, eh? You know, I come from a city where we can eat food from a different part of the world for every meal all week long if we want to, even Korean food…and no one thinks it's strange at all…most of us think it's totally normal, and we would get kind of bored eating the same food all the time…But, you know, I guess things are different in Korea, eh?" I try to offer him a graceful "out" from the conversation.

"Hmmm…yes," he says. His easy-listening music playing contradictorily in the background, he continues, "Yes, people can eat foreign food now…that is true…I don't like it though…," he finishes, turning back towards the computer beside him.

By now I've finished eating, so I stand up, get a cup of water from the cooler, and then wander over beside him. I know the source of his pain, and I empathize to a great degree, but I can't take it anymore. I just want to eat my lunch, check my email, and do my work without having to participate in this performance of petty fascism. A flash of images from Seoul's fascist history passes through my mind: Women stripped naked for wearing a skirt higher than their ankles, and men shaved bald in the street for having hair past their collars. I'm tired of being drawn into this petty lunchtime melodrama. I feel small for it, but I've decided I will finish this daily sparring for status.

"What's that you're listening to?" I ask him, already knowing the answer.

"Oh, this is some music by Roy Conniff…Do you know it?" he asks.

"Hmmm…yeah, I guess my GRANDMOTHER might have listened to that when I was a KID," I stress, then immediately turn to a sign on the wall above the terminal he is sitting at and read aloud in a bright, clear, broadcaster's voice that everyone in the room can hear, "Teachers are reminded not to smoke in the teacher's lounge. Please smoke OUTSIDE ONLY." I laugh a little bit, and say with a chuckle, "Se Ho… my friend…is this sign here just for YOU? I can't think of anyone else I've seen smoking around here…Did you get caught smoking in the lounge?"

Another teacher in the room, Ahrum, can't help but get in on this opportunity: "Yeah, that's for Se Ho…ha ha…" Ahrum is a Kyopo who, as the officially appointed manager of the instructors, is technically above both Se Ho and me. I look around at a few more faces that seem to enjoy this small revenge; none has the status to act out what I'm doing with Se Ho—even Ahrum doesn't.

Se Ho remains fixed on the computer screen in front of him and offers a weak laugh. I say, "Oh…sorry…I better get to work…," I walk back to my seat, wake up my laptop, and concentrate on my list of things to do. A while later, Se Ho slips out unnoticed along with the rest of the teachers' shift change. Although I'm not left feeling good about this entire escapade, he never comments on my food again.

* * * * *

Castells' seminal text in globalization theory promotes a reductionist view of technology: "all electronic processes…can be reduced to knowledge generation and information flows" (Castells, 2000, p. 409). Technology has its utilitarian purpose, making network society and globalization possible. For Castells, network is a matrix of governmentality. The aesthetic nature of global networks make possible a governing of the system, introducing profundity of systemic integrations previously unimaginable. The network displaces all else as the socioeconomic base. Network for Castells is "the space of flows"; as he describes, "the material organization of time-sharing social practices that work through flows" (Castells, 2000, p. 442).

Mobility reflects and/or constructs hierarchy in networks. The network unit of labor is the "networker": "The networker is the necessary agent of the network enterprise made possible by new information technologies" (Castells, 2000, p. 257). This laborer fits into one of three levels of network agency: Networkers "set up connections on their own initiative, and navigate the routes of the network enterprise," the "networked" are workers "online but without deciding when, how, why, or with whom," and "switched off workers" are "tied to their own specific tasks, defined by non-interactive, one-way instructions" (Castells, 2000, p. 260). The navigational autonomy of networkers thus endows them with an enviable position within the global system. Even so, in a Weberian slight of hand that substitutes status for class and hides power in institutional structures, Castells does not necessarily see networkers' agency as an endowment of power. Perhaps rightfully so, as Callon and Law (1995) point out, since agency is the ability to do things within the network array. Power, on the other hand, is something allocated by the network (Latour, 1986; Law 1992). For Castells, the global network is a new re-structuration of a familiar phenomena. Castells (2000) describes that power for individuals found in interactive, subjective identification (i.e., the mobility of the networker) granted certain positions in the network, in contrast with noninteractive systemic definition (i.e., the switched off workers).

The Story of Two Davids (Part 2, Scene 1)

The back office of Hoyah Academy in Seoul is a crowded little enclave of activity, and this afternoon is no exception. I'm in the payroll and administrative area. Desk space is at a premium, as the various "managers" have their territories here. It is a carefully arranged chessboard, with little room to negotiate one's way between squares. Among other things, this is where the school's lone LCD projector is kept. Since one of my classes is reading a version of *The Perfect Storm*, I've decided it would be a justifiable little treat for them to see the movie. The process is straightforward enough—I reserve the projector with the operations manager, Mr. Young (who looks after the web site and other technical aspects), and pick it up at the set time. I make my reservation for 4 pm the following day.

Across the passageway from the operations manager sits Johnny, the personnel and recruiting manager. He came from New York City to work as office help two years ago, about the time when I first visited Seoul. By persevering with the school and maintaining good relations with the owner, he got himself promoted up through the shallow hierarchy. He is the person who hired me on this return visit. I stop for a moment to say hello as I pick my way through the maze comprising the passageway to the exit.

"How you doing?" I ask.

"Good...good...," he says. "How are your classes going?" he asks me casually.

"Very well...," I say, as his phone rings. He picks it up with a " 'Scuse me," and I stand for a moment planning my next move. Behind him, in a dark corner (one might say, in the "farthest reaches" of the building), I see some bloodshot blue eyes looking at me.

"Hello," I say.

"Hi... I'm David," comes the reply.

"Oh, I'm Grant...," I tell him in turn. "You work here with Johnny?" I ask as we shake hands reaching out over Johnny's phone. I quickly appraise his appearance. Whether alcohol is the cause or not, he has the disheveled, swollen appearance of someone who drinks most of the time and has for a long time. The signifiers of his age have morphed into a decade and a half gray zone, anywhere from 50 to 65.

"Well, maybe it's more accurate to say I work 'beside' Johnny...I write class materials...," he explains the distinction. "I could actually work anywhere...," he says, gesturing towards his laptop, "...but I like having a regular desk to sit at." We chat a little more, and when he finds out I'm from Toronto, he tells me he is too. However, he's now a Korean citizen, having married a Korean woman many years ago. He tells me he first came to Korea in the 1970s. I tell him he's the first non-Asian Korean I've met. He tells me nonchalantly that his marriage didn't last though. I ask suspiciously why he is working at this school, expecting to hear that it keeps him busy in his retirement or some such thing. He tells me that it pays well compared to other schools he's worked at, and it's pretty easy work for him, especially since he doesn't have to deal with kids in this job. It's not the answer I was expecting to hear from someone his age, that he's a "career" English teacher. From his appearance I find it hard to imagine anyone willingly leaving their kids in his care for any reason, but I don't doubt that he has had ample experience. I excuse myself and head off to teach my first class of the day.

* * * * *

Andrew Barry suggests that much of what we do is geared towards surviving and reproducing technological society: "We live in a technological society to the extent that specific technologies dominate our sense of the kinds of problems that government and politics must address, and the solutions that we must adopt" (Barry, 2001, p. 2). This set of attitudes is governed through physical spatial mapping, and in the regulation of zones "formed through the circulation of technical practices and devices" (Barry, 2001, p. 3). Living in technological society requires one to acquire and constantly upgrade one's technical skills, providing such being a feature (even responsibility?) of government: "Technological inventiveness is a virtue...both rewarded and fostered" (Barry, 2001, p. 27). In Barry's terminology, political refers to "the ways in which artifacts, activities or practices become objects of contestation" (Barry, 2001, p. 6). Barry theorizes that institutions and other technologies and assemblages take on the appearance of intelligence insofar as the people who operate them are made invisible. This tends to reify the technologies alone as the infrastructure of international and global formations, but, in fact, such arrangements are highly dependent on the agency of a multitude of individuals (i.e., "technological society') in order to function, reproduce, and evolve. Altogether, as Khalideen (2008) explains, internationalization alone is not the same as global integration and global awareness. Intercultural conflict to some extent reveals the global shortcomings of an organization, as in a truly "global" organization there may be many cultural differences but few real cultural conflicts (Chan, 2008).

The Story of Two Davids (Part 2, Scene 2)

The next day at 3:45 p.m., I return to the back room to pick up the LCD projector before my 4 p.m. class. David's desk is noticeably empty. I have trouble finding Mr. Young at first, but when I do he immediately goes to the steel cabinet where the projector is kept. He takes out a key and turns the lock. He then casually turns the handle and pulls on the door. Nothing happens. He tries again, more aggressively. Still, the door refuses to open. He tries the key again, turning the lock a couple of times trying to loosen it, but the latch is obviously broken. He pulls at it from the top and shakes it, but apart from the dramatic effect, nothing happens. He looks at me, exasperated, smiling a little bit. I think about just ripping it open with my hands but then think the better of it when I realize they would probably charge me for any damage to the furniture. Mr. Young ducks into the next office and returns with the school's accountant, a forty-something "ajima" (aunt) whose name I was never told. She looks at me seriously, then looks at the recalcitrant cabinet, takes a step closer to it with Mr. Young right behind her, and reenacts all of his previous machinations. The effect is the same. By now, it's almost time for my class, so I tell Mr. Young I have to go and request him to bring it to me when they get it figured out.

My students are sitting expectantly in their seats, thinking they're going to see a movie. They groan their disapproval when I tell them there is a problem with the projector, but at that very moment Mr. Young heroically walks in and deposits the projector on the table. He turns and leaves just as quickly as he came, and I tell the class to keep quiet for a couple of minutes while I hook everything up. I take out my laptop, power it up, and hook up the monitor cable. I plug the projector into the nearest power outlet and try to turn it on. No power. I jiggle the plug, but it makes no difference. Spotting an outlet near the ceiling above a student's desk, I ask him to try the outlet. Still no power. I try another outlet on another wall, and still another, but I can't get the power light on the projector to glow. Perplexed, I decide to test the projector in the vacant classroom next door. The first outlet I try doesn't work either, but the second one I plug into finally gives me some satisfaction. "OK...," I think, "We'll just move in here..." I mentally prepare my speech to the class. When I reenter the classroom, the student who tried the plug above his desk happily announces he found a working outlet. We had tried only one of the two sockets, and apparently the upper one has some juice. We plug the power bar into the outlet, power up the projector, and start playing the DVD from my computer. All looks well...until I look up from the computer screen to the projected image. There is nothing on the screen. Almost as bad, I realize there are no speakers with the projector either, and it's almost impossible to hear the soundtrack. I decide I better restart my computer and let the auto-detect function find the display by itself. The technology has effectively hijacked the class, demanding all of our attention and stealing our time.

We wait a few minutes for the process to complete; I reopen the DVD player and try again. Nothing. By now we are almost twenty minutes into a two-hour session and I can only laugh at how badly things are going with the technology, but I find myself caught in the time paradox: I've already invested so much time in it that I can't allow myself to give up. I try a different media player application. When the second one doesn't work, I try changing players again. I try changing resolutions and frame rates in each player in turn, but I just can't figure out what the problem is. I ask the students whether any of them knows what's wrong, but to my surprise they don't. I finally decide to give up and that it just isn't worth it, but the students won't have it. They tell me they will gather around the computer and watch it. Since there are only ten students in the class, I decide they can actually do this to reasonable effect. I turn on the English subtitles and agree to let them do it. By now I really need a break anyway. With everyone's approval, I skip the "boring" first part of the movie and jump to the point where the fishermen are setting out to sea. I step back from the students hovering around the little 12-inch screen. "Technology spatializing itself...," I mentally note.

* * * * *

The spatialization of power appears as an early concern in the work of Michel Foucault, with allusions to the metaphor of network: "Power is employed and exercised through a net-like organization" (Foucault, 1980, p. 98). Although Foucault did not emphasize nor thoroughly develop a theory of network, his well-developed aesthetic disciplinary system of panopticism is indispensable to a sophisticated understanding of how aesthetics behave within the simulacra of the networked as a communications system. Foucault used the geographical term "archipelago" to describe a system "physically dispersed" but simultaneously covering "the entirety of a society" (Foucault, 1980, p. 68). The panoptic system is built on smaller, dispersed localized panopticisms, an archipelago of surveillance and discipline. Places within the system are panoptic islands within the broader archipelago social area. People move around and circulate power within these places. For Foucault, the power act of mobility is thus akin to circulation of individuals. The human subject is itself an apparatus—assembled within, by, and for institutions. He suggests: "The individual is an effect of power, and at the same time...it is the element of its articulation" (Foucault, 1980, p. 98). He continues: "Individuals are the vehicles of power, not its points of application" (Foucault, 1980, p. 98). Individuals thus transport and transmit power within the archipelago of panoptic systems. This leads to his terminology "Bio Power," in which "each individual has at his disposal a certain power, and for that very reason can also act as the vehicle for transmitting a wider power" (Foucault, 1980, p. 72). The individual mobilizes power within spatial formations in service to something greater than oneself. More succinctly, people act as conductors of power in the practice of communal interaction, and "space is fundamental in any form of communal life," ergo, "space is fundamental in any exercise of power" (Foucault, 1999, p. 140). The aspect of "communal life" provokes the terms "everyday life" and "individual agency" that Foucault's colleague Michel de Certeau is best known for and calls to mind James Carey's linking of communication to everyday practice and community.

The Story of Two Davids (Part 2, Scene 3)

The huge white canopy of the projector screen hovers mockingly above the little gathering of students. Why has the technology betrayed me? I consider grading papers behind them but instead take another quick look at my computer. In the dim light of the movie, I notice the F5 function button has something outlined in blue on it that looks like a screen. I reach through the little crowd of students and push the blue function key. The projector suddenly bursts into life, the movie abruptly cast onto the waiting screen. "Yeah!!" the students cheer this unexpected good fortune, a couple of them clapping their hands. Of course, the sound problem is still there, but they can actually read the subtitles now. Somewhat relieved, I settle in and watch the film with them. At appropriate points in the movie, I ask questions to compare the film with the reading. In fact, the story we read in class, a successful rescue of some boaters set in Canadian waters, is a minor part of the Hollywood film. This version focuses on American heroic bravado that ends up costing the entire crew's lives. The class ends up staying an extra five minutes to get to the end of the film, but they don't seem to mind. They quickly leave me to pack up the equipment once the credits start to roll, all technical anxieties forgotten. I drop off the projector back in Mr. Young's room. Neither he nor any of the rest of the usual gang of managers are there, but the cabinet door has been suspiciously left unlocked. I put the projector inside, shut it, and dash upstairs to my next classroom, a rather undramatic finish to the comedy of malfunctions the day has been. I don't show any films in class again.

* * * * *

Carey (1989) describes a seemingly de Certeauian actualization of "practice," which produces spatialization. Ritual in everyday life can be exemplified by how dance produces space, demarcating the borders of (or perhaps breaking through borders of) places through movement, or demonstrating through ritualized entrance and exit practices. Ritualized movement produces space as a knowable entity. Mobility makes it possible to know and make sense of the world through experience. In Carey's view, the separation of transport from communication "widens the range of reception while narrowing the range of distribution" (Carey, 1989, p. 136).

For Carey, power lies in the production and reproduction of goods and practices, in the performance of everyday life by everyday actors. He declares his ritual model is biased towards maintaining society across time, and towards representing beliefs that are shared among cultural actors. Ritual gives importance to communication as a process of "construction and maintenance of an ordered, meaningful cultural world that can serve as a control and container for human action" (Carey, 1989, p. 19). This might seem to represent humans as dronelike slaves to an "outside" order, but Carey is adamant that every human has a constructive role in bringing reality into existence through communication: "We first produce the world by symbolic work and then take up residence in the world we have produced" (Carey, 1989, p. 30). He asserts that only politics can control technology, emphasizing the agency of humans in relation to the things we create and operate. Thus he suggests that the role of the communications scholar should be "to examine the actual social process wherein significant symbolic forms are created, apprehended, and used" (Carey, 1989, p. 30). This is a very local type of directive in terms of methodology, emphasizing everyday places where patterns of ritual can be identified (e.g., building entrances: the entrance to a Catholic church, or an apartment building). However, the advent of globalization massively expands communications networks and exponentially increases the potential places of contact in which everyday ritualized activity may be performed between cultures.

The Story of Two Davids (Part 2, Scene 4)

I see David frequently as part of the background during my administrative dealings, but I don't really talk with him again until one day when he somewhat mysteriously appears in the staffroom. He is sitting where I often sit to eat lunch and to check my email. He has his laptop with him, and when I ask him what's up, he tells me he has been checking his email. I can see he doesn't really feel comfortable in this room. Since most of the teachers are in their early twenties, I imagine that most of them must look like children through his eyes. I take a seat at a terminal to his left and ask what he's been doing lately, wondering morbidly what his life is like. He tells me, "Just working...and last weekend I flew over a really great mountain range."

Interested in a possible outing, I ask, "Oh...you took a tour of some kind?"

He answers sort of secretively, "No, I flew myself...I'm a pilot..."

I halt for a moment but can't help asking, "Why are you working here if you're a pilot? I mean, how come you don't work for an airline or freight company or something? I'm sure it would pay better, no?"

"Ah, yeah, it would pay better, but I never got a commercial pilot's license, and I'm too old now," he says.

"Oh," I say apologetically, "That's interesting...so you just got your license for fun?" I continue, thinking I've found something admirable about my fellow Canadian.

"Well, not exactly," he answers, "I actually used to fly in the Korean air force." Now I'm totally perplexed, and my face shows it.

"How can that be?" I ask him, "How does a Toronto boy end up in the Korean military? Were you on some kind of a Canadian air force exchange program or something?"

He tells me, "No, I was in the Korean air force...It's kind of a long story...but my father spent some time working in Chile in the 70s [he looks at me to see if the statement has an impact, which it does] and South Korea bought some aircraft from Chile at that time, and I came with the aircraft as a pilot and trainer...," he finishes with an air of mystery.

I take a moment to digest the revelation of his fascist legacy. I think about how Allende was bombed by the Chilean air force during the 1973 coup d'état. Did this wreck of a man train those pilots or, worse, fly one of those planes? Did his father sell those planes to Pinochet? I decide not to ask. He's already answered my questions, and what difference does it make now, thirty-two years later, anyway? I look at his puffy, drawn out face and imagine how great it must have felt to be him in the 1970s, firmly allied with America's puppet capitalist dictatorships as a uniformed technical expert...such status and power...part of the machinery of death, and no doubt employed to deliver. I think about the many children of Chilean exiles who I became friends with back in Toronto. I think about how hard winning the end of fascism was in South Korea, how most of my students don't know that South Korea used to be a dictatorship and how most older Koreans I know simply refuse to talk about it.

"Hmmm...," I say, "I don't think a lot of people around here know what that really means...but I do," I tell him evenly, emotionlessly.

"Uh, yeah, well, I just fly for fun now, a small plane, on the weekends...," he says uncomfortably, rising from the chair. "I better get back to work downstairs...I came up here because the email wasn't working down there...see you later...," he says as if he's done something wrong, exiting through the door behind his chair. I sit quietly and watch the door close and look at the young Kyopo teacher sitting beside me, typing furiously at her terminal. She seems oblivious to what's happening in the room, probably unaware a conversation just took place behind her back. I pause and reflect for a moment on how his story must weave into her history, and mine, and back and forth again, then, overwhelmed by the enormity of the connections I'm drawing in my head, I empty my mind and turn to my own terminal. I launch the browser and check my email. I often see David working on his laptop at his desk in a dark corner downstairs, but I never see him in the staffroom again. We always nod hello but never actually speak to each other again.

* * * * *

Globalization of the performative arena is an important concern of David Morley. Morley and Robins (1995, p. 1) state that transportation and communications networks provide "the crucial and permeable boundaries of our age." This statement is based on an early description of network by Manuel Castells, primarily focused on economic production but problematized by the issue of cultural mobility and human migration. This produces a delocalized world order articulated around a relatively small number of production centers: "Particular localities and cities are drawn into the logic of transnational networks" (Morley and Robins, 1995, p. 73). Place is thus a nodelike entity that trades and communicates with several other productive places, thereby creating networks. The emphasis in Morley and Robin's work is on the local, and the way in which concepts of belonging are being forced to reconcile with the manifestation of aesthetics and performances once considered "other." The identification of local peoples and their rituals of everyday life are altered by the appearance of signifiers (i.e., bodies and practices as well as material goods) that can no longer be considered apart from the local places they have migrated to: "the stranger, the foreigner, is not only among us, but also inside us" (Morley and Robins, 1995, p. 25). Thus mobility in Morley and Robins' writing is the circulation of people, signifiers, and goods from one place to another.

Morley's concept of mobility as transport of the self and one's home environment contributes an important element to understanding the individual in this milieu. For Morley, mobility is a taking of home with oneself, and an "escape from geographical location" (Morley, 2000, p. 150). Inversely, mobility of content can mean an invasion of unwanted symbols into a home territory. Personal mobility (as in transport of one's self) is power, while "enforced immobility" (Morley, 2000, p. 158) serves as a demonstration of discrimination. Working with the premise of "the inside outsider," Morley problematizes notions of clear distinctions between inside and outside, belonging and foreignness. An understanding emerges from his work of a complex, ever-changing system of interdependencies. In the present, the survival of local regions depends largely on their close links to the global electronic conduits of capital (Lee, 2008, p. 3).

Mattelart explains that creating networks of global interdependencies furthered the enlightenment project in terms of communications by "freeing" flows (free in the laissez-faire sense of the word), influencing language, the emergence of communication through signs, and the imposition of standardization of that communication to make it more widespread. In a Liberal sense, it created an international division of labor, unified trade areas of circulation, formulated the symbolic representations of the industrial "nation state," and brought about "world time." Since then, global interdependence has evolved enormously, and there has been an "awakening of planetary consciousness" (Mattelart, 2000, p. 66), but this global consciousness is dominated by standardized corporate information structures. Individuals are profoundly interdependent in this global system, in which inter-reliance on anonymous networks of production and consumption becomes taken for granted. But just as Heidegger pointed out, connectivity doesn't in itself bring nearness or authentic understanding.

Genetic Superiority in Global Network (Scene 1)

"Sometimes you just have to have one...," I hear myself say. My friend Sean and I are sitting at a coffee shop in Seoul's COEX mall, where we have just checked in on the status of his laptop in for repair at the Apple store. We're discussing eating American food—hamburgers in particular. We agree that when living abroad, from time to time one suddenly craves the most mundane familiarity in food. Where to get a decent hamburger...? We dismiss the obvious (McDonald's, KFC, Burger King), agreeing that we want a "real" burger (not fast food). Noticing a TGI Friday's restaurant down the hall, we agree they'll have the type of burger we're looking for. Although less crowded than most other restaurants in the mall, the only space available without a wait is at the bar. We're fine with that and take our dimly lit seats alongside some beer-swilling patrons. We busy ourselves ordering and chat a bit about the decor and other patrons. Our food soon arrives. Our dining proceeds as nondescriptly as one might expect for two good friends. Part way through, the two seats to Sean's left become occupied by two young men, their short hair and squared physiques creating an obvious portrait of American soldiers out on the town. One seems to have a Virginia accent. Sean becomes interested in our new neighbors, his journalistic imagination running wild. He tells me he wants to ask them some questions about their experiences of being stationed in Seoul. He wants to know where they're from in the United States, why they chose to go to COEX mall, why they are eating at TGI Friday's, what they think about Korea in general. Then Sean falls silent himself, eavesdropping on their dialog and trying to figure out how to insinuate himself into their conversational space. I remain silent too, trying to leave him the room he needs to do his investigating. In the quiet between us, I become suddenly aware of another conversation going on along the bar to my immediate right.

Without meaning to, I realize I've been listening to the ramblings of a beer-glass pundit. Sitting beside me is a man I later come to know as "Sarge," his body turned towards a woman to his right, giving me his back. He's been on a diatribe for some time, basically ranting about how bad and stupid and fucked-up the rest of the world is, and how superior and great the United States and the Americans are, even blacks (although not as good as *normal* Americans). "Wow!" I think to myself, "It's just like being in Champaign, Illinois…this guy must listen to a lot of talk radio…" I look over his shoulder and see a very average white woman with short, brown hair, probably in her late twenties or early thirties, nod absently, stating "Uh huh…" in compliance with his quasi-scientific sounding racism. At first, I think maybe she's a soldier too, and thus more or less forced to agree with Sarge, but I begin to see from her behavior that her interest is not so much military as personal. While not overtly sexual, she's too flirtatious to be a soldier. I decide she must be a support worker on the base. Maybe a secretary? Kitchen worker? Who knows. Whatever the profession, her unwavering eye contact signals that she's definitely into Sarge, and not for his political sophistication, which I start to think is something she's willing to put up with to be with him.

My illusion is quickly shattered. She abruptly participates in the conversation by espousing some of her own prepackaged xenophobic tropes, enframing the United States as constantly under attack, victimized by a cruel, ignorant world. "Why can't they understand we're doing what's best for the whole world…?" she concludes with a query to her seemingly omnipotent companion. The caricature is so perfect that I'm beside myself with amusement at this point, staring down at the plate in front of me with an incredulous half smile on my face, oblivious to Sean and the soldiers next to him. I look up to check on the two Korean men behind the bar, to see if they've been audience to this geopolitical melodrama, but they are busy working with other customers around the corner. I can hardly believe this is a serious conversation, and I look quickly past the woman, to see if I'm the only one who is so shocked by their ignorance. I see two more soldiers who, by their body language, appear to be with Sarge and the woman, but who seem to be off in their own beer-induced fantasy.

Sarge suddenly retorts by going into an account of his understanding of global economics, which is basically to say that the United States pays for the entire world, and poor countries should be cursed and punished for not being able to pay their debts. "But you know, it's not really their fault...," Sarge states in a burst of fatherly compassion. In a hushed, caring tone, he confides, "They're poor because they're genetically programmed to be...," I snort my incredulity into the plate in front of me. I turn and look at Sean, my eyes wide with astonishment.

"Did you hear that?" I ask him.

"Hear what?" Sean replies. He's still waiting for an opportunity to break into the conversation to his left. I quickly look around at the rest of the room and haphazardly note that everyone I see sitting in the booths is Asian—in stark contrast, for whatever reason, the bar is populated entirely by Westerners.

"I gotta listen to more of this...," I stammer. "Just give me a minute...," I say, turning my attention back to Sarge.

"It's true...," I hear him affirm, "We're all *de*-evolving now...You know, since the races have been corrupted...," he explains, with a special emphasis on "de." "So all the poor genes are mixed with the other genes, and we're all *de*-evolving because of that...and it's too late to stop it...," he goes on pensively, lamenting his own fate as an inferior white male, a victim of history. The woman beside him continues her silent nodding of agreement.

One of the other soldiers in their party suddenly interjects: "They're poor because they're not FUCKING AMERICAN!!" he says jovially, to the cheers and laughter of the rest of the small group. Their party is about ready to leave now. The man beside me pays the bill with his credit card.

"Thanks Sarge," the other two soldiers say as a chorus.

"Exactly what I needed!" says one happily. My eyes follow them as they walk towards the door. I muse to myself that this is a man who, under certain circumstances, decides who will live and who will die. I turn back to Sean, who tells me he just doesn't seem to be able to find a point to jump into the conversation beside him.

"It's OK," I say, "sometimes things go that way..." We pay our own bill and follow the tracks of Sarge and his crew out the door.

* * * * *

Vega and Vessuri (2008) have done a good job explaining how national innovation systems are designed and funded (or not), and how because of it there is little chance of any nation sneaking up from behind to become a player in the global innovation economy. Things are ordered this way because the world is structured to be such. However, participation in the global economic network also necessitates the dimension of participating in a global "moral economy" (to use James Hay's term). Relations of labor and capital, reconfigurations of everyday life and ritual, and fundamental issues of identity and performativity conspire to challenge individuals in global networks to make sense of, reinvent where necessary, and enact their morals and ethics in an entirely new context. Hardt and Negri explore the complexities of this individual/network relationship, theorizing the network as fundamental to the contemporary conception of power: "Power can be constituted by a whole series of powers that regulate themselves and arrange themselves in networks" (Hardt and Negri, 2000, p. 162). The network is the "organizational model of production" (p. 295), deterritorialized in a traditional Euclidian sense but comprised of "sites" coordinated via information networks and the mobility of capital from one site to another. Network has come to constitute all space: "There is no longer a place that can be recognized as outside" (p. 211). In this theory of network, place is, in fact, negated entirely: "In Empire, no subjectivity is outside, and all places have been subsumed in a general 'non-place'" (p. 253).

Mobility for Hardt and Negri is defined primarily in terms of goods and labor. Controlled mobility has always been a requirement of capitalist accumulation, but contemporary methods of production coupled with the desirability of mobility to improve worker's living standards on individual levels make the present situation unprecedented. The people constituting the global labor pool form a constantly shifting "multitude" (Hardt and Negri, 2000, p. 397). But contemporary mobility continually breaks the traditional boundaries and is less containable, and thereby the multitude "reappropriates space and constitutes itself as an active subject" (p. 397). Contemporary production happens in the space of movement and depends on migration of both workers and capital.

As for the issue of power, the global situation is theorized in terms of a U.S.-led network of capitalist empire: "Empire can only be conceived as a universal republic, a network of powers and counter powers structured in a boundless and inclusive architecture" (Hardt and Negri, 2000, p. 166). Hardt and Negri's work builds on Virno's (2004) notion of multitude but seems to use the language of Actor-Network Theory in discussing power as network, with a pronounced absence of dualistic "inside/outside" distinctions, with the appearance of authority as aesthetic "singularity," and with an overall concern with ontology. A decade after Latour's exposition of Machiavellian "machinations" (1988c), Hardt and Negri described a networked system of power within which individual actors produce and contest hegemonic Machiavellian political machines. Speculations of the origin of their terminology aside, they describe that the multitude creates "constellations of powerful singularities" (Hardt and Negri, 2000, p. 61), and thus the performance of every individual in the multitude profoundly contributes to the maintenance and expansion of the global network. The epitome of this process of creating singularities, the sovereignty of the United States is "the result of the productive synergies of the multitude" (Hardt and Negri, 2000, p. 164). For Hardt and Negri, only the actions of individuals already within this sophisticated global network can alter its conditions and bring relative liberty (or perhaps simply lessen the conditions of oppression) for those who suffer the consequences of globalization. The method of governing individuals in this context returns this review to the work of Foucault through a reading by Jeremy Packer.

Genetic Superiority in Global Network (Scene 2)

"Can you believe it? He actually said 'they're poor because they're genetically programmed to be'...I mean, he actually used the word '*de*-evolving'...is that even a word?? Man, and this is someone who tells other people who to kill!" To myself I reflect, "No damn wonder Se Ho hates hamburgers..." Sean and I are now at the movie theater about to see the new version of *War of the Worlds*, described on YahooMovies.com as "The extraordinary battle for the future of humankind through the eyes of one American family fighting to survive it." Sarge isn't in the drama we're about to see, but his stand-in Tom Cruise is. The Orson Wells' classic retold is about an alien invasion of the earth. "Foreigner watching a film about alien invasion in Korea...," I mentally note. I also note that the English soundtrack remains intact, though it is subtitled in Korean. In the end, in spite of their technological superiority, the invading aliens are beaten by their inability to live on earth. Their alien "genetic program" lacks the ability to fight earthly diseases, causing them to succumb to earthling germs and viruses and such. They are contaminated en masse, exterminated by unwitting microscopic organisms in spite of their overwhelming technological dominance. Bereft of their users, the technologies that signify their superiority collapse into useless piles of junk, not unlike the state of Sean's laptop sitting in the repair shop. But life goes on, the foibles of the individual earthly character highlighted by their insignificance in the context of an intergalactic battle. As the credits start to roll, something suddenly dawns on me. Thinking through the myriad idiosyncrasies and eruptions of petty oppressions and resistances, interpreted through the narrative of *War of the Worlds*, the irony of our best intentions becomes crystal clear: Everywhere we go, regardless of where it is in the world, we participate in the performed reproduction of this war—as victim and as hero—righteously, willingly, shamefully, unwillingly, and even more often than not, obliviously. It involves places, movies, hamburgers, beers, laptops and cell phones, and everything else we take for granted in everyday life. But over all, being "global" doesn't in any way entail being any closer together. Ironically, it's about the people closest to my own mundane everyday existence—my friend Sean, my brother and family, my close friends—that I write the least. "I guess the aliens were just 'genetically programmed' to lose, eh?" I joke to Sean as we exit the theater. "No man, they lost because THEY'RE NOT FUCKING AMERICAN!!" he shouts with a big smile.

* * * * *

Packer uses the term "navigation" to discuss the relationship of mobility to power (Packer, 2003, p. 135). If "individuals are the vehicles of power," then "personal mobility must therefore be seen as an act of power" (Packer, 2003, p. 136). Governance is manifested as control over how individuals as vehicles navigate their way through the world. Apart from physically preventative or restraining apparatuses, the ideological construct "safety" provides justification for a disciplinary regime of normative behavior that protects mobility as a political, cultural, and economic good (Packer, 2003, p. 136). Safety thereby provides the context legitimating pretexts for normalizing modes of governance and self-governance. Thinking back to Morley's "inside outsider," the desire for safety from intrusion by "others" one has learned to fear is a strong motivation to adopt the discourse of safety. Packer works with the notion that "agency is organized through the control of mobility" (Grossberg in Packer, 2003, p. 141). The potential for interloping "others" is systemically reduced and policed so as to appear a natural part of everyday life. Hardt and Negri's "multitude" is thus kept in its "placeless place" by an ideological and physical apparatus that presents itself as self-evident for the benefit of those it controls. Of course, one is hard pressed to think of safety in itself as undesirable, but Packer outlines how the construction and appropriation of individual fears are accomplished through one of the three modes of objectification described by Foucault: "the self turning itself into a subject" (in Packer, 2003, p. 154). Persuading and conditioning individuals to monitor and police their own mobility is the goal of this particular strand of objectification. This independence of movement along with new regimes of personal mobility necessitate a new terminology: auto-mobility.

According to Edensor (2004), "auto-mobility" is one explanation of how national identity persists and continues to grow in reach in spite of the rhetoric of "borderless" globalization. He explains that although understood as inherently fluid, auto-mobility is always contextualized as situated within complex matrices (Edensor, 2004). This is obviously in keeping with the predominant understanding of network as conduit populated by transporters of power, but Edensor is concerned with pointing out the resilience and rebirth of cultural belonging through the performance of mobility in the context of globalization. Performances of auto-mobility "sustain both conscious and unreflexive impressions of national belonging" (Edensor, 2004, p. 111). The embodiment of nationalistic codes enable citizens to act like "themselves" (i.e., enact the performances and logics of their learned cultural scripts) regardless of location. Edensor suggests that performative places within geographies continue to provoke eruptions of identities and belongings in spite of the suggestion that it could be otherwise. Carey might question why this would be remarkable at all, given that ritual is the process of making life meaningful in the first place, and such processes will always prevail. However, following from a Foucaultian understanding of governmentality, it might be said that it is not the fact that people perform rituals that is problematic. Rather, one might take issue with the detail that power insinuates itself into ritualized activity and thereby appropriates and even hijacks auto-mobility. Edensor's work thus leads back to the complexity of the relationship between individuals' performative capacity to be productive yet mobile and unchanging, and the contexts in which we seemingly inescapably live and move (and perhaps all the more inescapable as global networks continue to expand).

Dreaming Insecurity

I find myself along with two others taken hostage by a gunman (who I somehow know, but just can't quite place). At first, I don't take it seriously. In fact, I think it's a joke. But when he points the gun at me, I realize he's serious about killing us. I try to remain rational, beating down a flutter of anxiety. For some unknown reason, I decide I'm faced with the either/or choice of fighting over the gun, or tricking him into letting me go to the bathroom. Somehow, playing on the feelings of friendship we have between us, I convince him to let me leave for a minute by promising to return. Once out of the room, I run away, calling 911 on my cell phone.

I'm suddenly in a crowded school in the northeast part of the city (what city I don't exactly know). I describe the situation to the operator, but the signal drops before I can give my exact location. I walk around the school trying to find good signal strength somewhere. I run into some friends in the school who seem to be shopping in the stores that I suddenly realize are lining the halls. When I explain what I'm doing, they are surprised the person is holding hostages anywhere in this mall atmosphere. I leave them and continue looking for a strong wireless signal. I finally find a place my phone will connect and call 911 again to give my location (specifically, Building 925, DFR School).

Now I'm awake, puzzled by the dream, and most specifically by the number 925 and initials DFR. I get out of bed, walk over to my computer, and type "925 DFR" in the awaiting Google search bar. Google fails to return anything significant to me. I brew some coffee and begin my regular, mundane morning routine, reassured by my ritual, secure in the "real" world. Connectivity brings me a feeling of security.

* * * * *

James Hay writes that definition as an accomplishment entails both "the production of meaning" and "spatial production" (Hay, 1996, pp. 359–360). Definition is a "process of marking, demarcating, tracing, and connection points" (Hay, 1996, pp. 359–360). Landscape as a Certauian technique refers to "a field of activity and multiform practice that surround analysts and their object of analysis" (Hay, 1996, p. 360). Although it might appear relatively easy to demarcate or map a space from a privileged vantage point, it is more difficult to locate the people being theorized within that space. Although space is traditionally treated as a much more static entity, people in everyday life move around spatially and also change identification. Hay lays out this problematic in a summarized discourse among scholars concerned with "locating" audience members:

> Radway proposed [locating the slippery audience] through an "ethnography" that...could more effectively consider the practices of everyday life "as they are actively, discontinuously, even contradictorily pieced together by historical subjects themselves as they move nomadically via disparate associations and locations through day-to-day existence. (Hay, 1996, p. 363)

While providing a good summary of the problem of mutating identity of individuals in everyday life, Radways' ethnographic solution was rebutted by Grossberg's suggestion that subjects' mobilities have to be taken into account. The question, then, becomes "when is one not an audience?" (Hay, 1996, p. 363). This very problematic seems to be the concern of Virno, in theorizing "the multitude" primarily as an audience for the staging of events. Virno might suggest that one is not an audience when one is performing for the multitude—in an Actor-Network Theory mindset, one might say one is not an audience when one is translating.[6] Regardless of such emphasis on individual shape-shifting, the processes of constituting and reconstituting everyday life are generally taken to be somewhat consistent, and they happen in time and space. Part of what's missing is what people actually do with media in their everyday lives—what do we in our everyday realities do with our global network?

Hay writes about everyday life as a "spatially constituted field of practice" (Hay, 1996, p. 363) that refers to "the 'kaleidoscope' of social practice and lived experience that exceed attempts to fix them as social structure, but also to the transience of social subjects amidst the spatial organization of the social world" (Hay, 1996, p. 364). The problematic of definition as practice is found in how it is "continually resituated yet territorializing" (1996, p. 366). Hay suggests that a terminology of "landscape" can encompass the contradictions and sophisticated actualities enacted among individuals, and between individuals and their environments:

> Landscape is that which already is spatially organized but which is continually traversed and gradually reconfigured. And as such, it discourages a simple opposition between the "geographic" and the nomadic or temporal. (Hay, 1996, pp. 367–368)

6 For an ANT definition of "translation", see Glossary, p. 177.

The materiality of landscape depends on the *practice* of its materiality, even while the materiality of landscape conditions and gives aesthetic form to society as a structure to occupy. As such, "the speaking subject" should not be privileged above the "landscape of material and discursive practice" (Hay, 1996, p. 369), an idea that clearly resonates with Callon and Law's definition of actor as an entity struggling for definition through the confusion of network distributions that are thrust upon it.[7] Hay suggests that Massey's "power geometry" offers a way to link social groups and people with specific places, at the same time reducing the emphasis on only human actors by paying attention to their landscape. Working with this premise, Hay later turns attention to the complex relationship between mobility and network.

No Connection: A National Failure?

Montreal is always a contradiction to me. It *should* be great, I *know* it is great, but I just never really seem to *experience* Montreal as great. It's just that nothing great ever seems to happen to me or around me in Montreal. Although I know it too is part of "my" country, I always feel like a visitor here. It's my second sunny day in the city, and I decide I need to check my email. I take my laptop and walk down St. Catherine's, seeking a coffee shop with a hotspot. After trekking half an hour in the hot afternoon sun, I decide to just park my butt inside an air-conditioned café. Out of habit, I open my browser and look at my Internet connection menu to see whether I can pirate a signal from somewhere. No luck—the digital air is dry as a bone. My browser produces an error message telling me it couldn't load the Toronto Star homepage because there is no connection. "No connection…," I muse pensively to myself. Looking around the café, I realize this is a francophone hangout. "Nes pas conexion…," I say to myself in amusement, pretending I can speak a few words of French. The woman next to me asks something in French, which I take to mean she wants the unoccupied chair at my table. I gesture my acquiescence, then, sans wireless, turn my attention to proofing a journal article on globalization and culture.

* * * * *

Hay explains that media links network and mobility insofar as technological appliances function as both "site and network" (Hay, 2001, p. 213) much

7 For an ANT definition of "actor", see Glossary, p. 169.

like music links dancers to spaces, while participating in an organic relationship with individuals and their landscape. From the network side, technology as a site comprises and is comprised of an "assemblage of practices" (Hay, 2001, p. 211), forming a "sociospatial problematic" (p. 212). Mobile technology functions in much the same way as less transportable devices. However, as if in conversation with Barry's notion of technological society (2001), Hay points out that mobility becomes "a necessary technique for living" within the networked context. Access as an achievement and mobile technologies as allies are the arbiters (as in the technical definition: protocols) and measures of freedom and power in this understanding of networked landscaping. Governmentality takes on the characteristics of policing auto-mobility (per Packer above). Hay's description of "mattering" describes governance in this situation—devices are complexly interconnected, the device being an object that both governs and is governed. With television as his case study, he describes that TV as a network of sites is governed "through the production of intersecting, interdependent networks from infinitesimal sites along the network" (Hay, 2001, p. 217). This more sophisticated description furthers Foucault's underdeveloped notion of the archipelago of panopticisms that constitute society. In keeping with a Foucaultian analysis of power, technologies are "assemblage of practices" that are "dependent on and instrumentalized through a broad array of practices and technologies" (p. 211). Technology constitutes a site of converging "practices and technical competence" (p. 212) with the objects and environments themselves.

The work of Hay brings into dialog the extremes of Castells' theorized network structure with a Foucaultian analysis of power, acknowledging the importance of both while refusing to privilege one over the other. Equally important, Hay's elaboration of "landscape" offers a language to help guide ethnography in this present media circumstance: Investigating global mobility through the spatializing agents of wireless mobile technology (the cell phone and wireless laptop) that serve as actants in the global network, elaborating the performance of the actor-networks (including human actors) in which they are allied and, ultimately, describing the localized landscapes that they spatialize as practices, constructing site and network within a broader global context of cultural, national, and local territories and places.

CHAPTER THREE

Technological Mobility and Cultural Practice

"Being" Canadian: Hide the Cell Phone (Scene 1)

The Starbucks on the corner of College and Yonge is a well-known establishment, directly across from my favorite discount shopping store. I enter through the east doors, the blustery spring afternoon wind gusting for a moment. I walk up to the counter. Strangely, there is no line just now. A white woman in her early twenties with long curly black hair asks if she can help me.

"Do you have wireless Internet here?" I ask.

"No," she says apologetically.

"Do you happen to know anywhere nearby with wireless?" I ask. She turns for a moment and asks her younger, geeky-looking male colleague whether he knows anywhere. He looks at her, then me, then meekly squeaks out a "No…"

"You don't know?" I verify. They both shake their heads no. The woman I started with sounds a conciliatory note, telling me about a cybercafe down the street.

I say, "OK, thanks…," and turn towards the door from whence I came.

"Have a good lif…," she chokes on the word, stammers, "… a good luck," and manages to finish as I walk towards the door laughing.

"Yeah…sure…"

I walk north on the east side of Yonge Street, towards the Second Cup coffee shop less than a block away. My brother's first job in Toronto was at a Second Cup, so the chain always holds a special place in my heart. I see a sign in their window advertising that free wireless is available inside. I verify the information at the counter, then order a latte. I choose a strategic seat and spatialize myself with my technology next to the Yonge Street sidewalk window. Their start page loads, telling me the Second Cup is "proudly Canadian." "You go get them damn American Starbucks!!" I cheerlead sarcastically in my head.

As I check my email and sign into MSN Messenger, I notice through casual observation what my friend Celiany said is true—people really don't seem to use cell phones in public here. Almost no one walking on the street is talking on the phone. I finally see one woman carrying a very old one with the antenna out. I peer long and hard across the street to the other side and look around inside the coffee shop. Inside, about half the customers use laptops, and it seems like half the other customers are practicing English. I don't see a single cell phone in the place. Outside, I decide the ratio must be about one in forty people using a cell phone. I hadn't noticed this before, but compared with my observations in other cities, I can certainly verify Celiany's conclusion. "Maybe it's too cold?...Interesting...," I mentally note to myself, deciding this issue requires further attention.

Figure 3-1. Nationalism Online. The Second Cup—A coffee house chain uses nationalism against its competitor Starbucks Coffee, promoting itself as "Proudly Canadian."

* * * * *

In light of poststructural and postmodern contributions to media studies, I propose modifying McLuhan's (1964) famous phrase to say the medium is *a* message, one of potentially numerous aspects of a given medium that include ritualistic and cultural scripts as well as media content. With multiple origins, the result is activity in the world that can be identified as affiliated with (if not entirely caused by) particular devices.

James Carey described that ritualized aspects of communication bind society through time, bringing a sense of constancy that unifies culture through the ages: "A ritual view of communication is...directed toward the maintenance of society in time" and "the representation of shared beliefs" (Carey, 1989, p.18). Ritual gives importance to communication as a process of "construction and maintenance of an ordered, meaningful cultural world that can serve as a control and container for human action" (Carey, 1989, p.19). In a statement that resonates with Charland's technological nationalism and the creation of imagined cultural territory, Carey theorized human activity: "We first produce the world by symbolic work and then take up residence in the world we have produced" (Carey, 1989, p.30). Similar to de Certeau (1984), Carey suggests that the study of communication should be localized, concerned with examining "the actual social process wherein significant symbolic forms are created, apprehended, and used" (Carey, 1989, p.30). Carey's theory situates media devices within a web of activity and meaning, within a process of creation, apprehension, and use. Technologies are part of localized symbolic work. Mobile technology is concerned with "taking up residence" in the world territorially ordered by nationalisms and other culturally significant spatially binding symbols.

"Being" Canadian: Hide the Cell Phone (Scene 2)

Being totally subterranean in the downtown area, it's a rare occasion that anyone can use a cell phone on the Toronto subway. However, as we emerge out of the tunnel and cross the bridge spanning the Don Valley, I decide to take the opportunity to check my voice mail. I self-consciously pull out my phone, seeming to be the only one on the car who would dare. As the late afternoon sun breaks through the windows, I listen to a new message from a friend who I am meeting later. My gaze passes from one dartingly averted set of eyes to another, making a round of the area within eye contact. By the time I finish listening, we're almost entering the tunnel on the other side, so I self-consciously fold up my phone and slip it into my pocket again.

* * * * *

Latour considers an actant anything—individual person or thing, crowd, figurative or nonfigurative entity—that exerts a force in terms of an act or an exertion of agency. Actants gain strength by joining force either willingly or coercively with other actants in alliances, basically representing themselves as a single hegemon. Actants work to make irreversible change, usually in unequal relations. In a process of programmatic reification, one actant learns from others, then endeavors to replicate what it has learned elsewhere, to program other actants the same way (Latour, 1988a). The dominating actant/force is so because of its ability to translate the other actants' language to the broader network of relations, either with the consent of others or by silencing them. There is always a relative power imbalance in this, but even so, it is not possible to define the location of actants. Latour elaborates: "We can only say that some locate and others are located" (1988b, p.164). Space and time can locate only those actants that are in submission to the hegemony of another actant, meaning time and space as descriptive frameworks are in negotiation with the dominating force.

Machines often perform and improve on human skills, while humans can perform machinations and resist individuation in the context of administrative apparatuses (i.e., the police, army, etc.). The issue in terms of the forfeiture of power is in translation "from one repertoire to a more durable one" (Latour, 1988a, p.306). This is an inscription or encoding that occurs in the translation from body to machine, as in when a machine is implemented to perform a task formerly done by humans. "Prescription" is whatever is presupposed from "transcribed actors and authors" (i.e., those forces who are part of the alliance of actants), including presuppositions encoded in machines (p.306). Individuals can become part of the alliance by being "inscribed" by the actant in a process I understand to be similar to Althusser's (1971) description of interpellation. "Des-inscription" is breaking from prescribed behavior, while "subscription" is acquiescing to it (Latour, 1988a, p.307). "Pre-inscription" is everything that prepares the scene for articulation (p.307). Continuing with his own act of translation, Latour describes that "sociologism" is one's ability to read the scripts of nonhuman actors, while "technologism" is the ability of humans to read their own behavioral scripts prescribed by technology (pp.307–308).

Machines are "lieutenants" that "hold the places and the roles delegated to them" (Latour, 1988a, pp.308–309). Machines are given birth through an elaborate network of alliances, but once put in place, they enact in the everyday world the behavior programmed into them: "what defines our social relations is, for the most part, prescribed back to us by nonhumans" (p.310). So what is similar and different about what wireless devices prescribe back to us?

What is described above is generalizably applicable to how any encounter with any technology might be understood. However, differences between technologies can be more easily teased out of the notions of network and "technology as lieutenants." More important in terms of human relationships is who commands these lieutenants. As a translator of that power, almost any user may become relatively stronger by the hegemonic potential of commanding an actant in alliance with the global network. On the other hand, mediated communication is also rife with all the social customs, rituals, and norms afforded every interpersonal exchange. Media practices are more akin to (re)negotiations or (re)affirmations of alliances and cannot be counted on as hegemonic machinations.

"Being" Canadian: Hide the Cell Phone (Scene 3)

"So... why do YOU hide your cell phone?" I ask Walter. We're standing on Spadina Avenue, just a few blocks from his house.

"Well...I don't know if I would say I HIDE it, that might be a little strong...and kind of expresses intent, which I don't know if it's really there...but I just usually keep it put away..."

"Yeah, OK, so why do you keep it put away then?" I refocus on the phenomenon.

"Oh, well, I'm not really sure...Why do you ask?" he flips the interrogation.

"Because a friend of mine noticed that Canadians don't really seem to use their cell phones in public, and I kind of think it's true, but whenever I ask someone about it they usually do have one with them...they just don't seem to ever like to use them. And I know YOU have one, but I don't really ever see you use it or even check it..."

"Hmmm...that IS kind of interesting...," he says, and I can see him switch to his analytical mind. "Hmmm...maybe it has something to do with the class dynamic...you know, Canada is so overwhelmingly middle-class and colonial..."; ever the sociologist, he begins to theorize the phenomenon in general.

"Well, at first I thought maybe it was just the weather, you know, like that it's just too cold to hold the phone to your head all the time...," I explain.

"Yeah, that's an interesting point," he nods.

"But then, I have been watching people both on cold days and warmer days, when it shouldn't bother anyone to hold their phone, and I don't really see any difference...," I say.

"Sure, yeah, but maybe, you know, people are just acclimatized to not use their phones then, you know, because they get used to not using them in cold weather and just aren't in the habit of using them in the warmer weather either," he concludes logically.

"Well, but there's a couple of problems with that...," I tell him. "Like, people don't seem to like to use them in indoors places either, like on the bus or in restaurants or lobbies, and then...you would think people would just use headsets more often if the cold was really an issue..."

"Hmmm...yeah, you're probably right...," he says.

"So what about YOU? I mean, personally...why don't you use YOURS more?" I ask again. "Is it your phone plan? Like, is it too expensive or something?"

"Well, you know, it's not cheap...but no, I don't really think about the price too much when I use it. Like, I pay the monthly plan anyway...," I nod in agreement, having already checked and found the rate plans to be basically the same as in the United States. "Hmmm...well, I think it's probably an issue of class and stigma, you know, like the cell phone got classed as a luxury item, and we try to keep it protected or something like that...," he abstracts his own experience and generalizes again, seeming to be unwilling to confront the question in the context of his own everyday experience.

"Interesting...," I say. "So you think you keep it put away to preserve some sense of status attached to it?" I confirm.

"Well, I don't know...I just keep it put away!" he says with a smile and a laugh.

* * * * *

De Certeau moves one's attention from the grand schemes of civilizations and nations discussed by Innis and Charland to a much more intimate scale. He suggests that only by looking at local practices can one discern the ethics, aesthetics, and struggles that constitute everyday life within the broader contextualizing socioeconomic schematic. Even so, he cautions that it can be hard to discover poiesis (one assumes a reference to Heidegger here) in everyday life because it happens in the space of "production" and in the guise of consumption (i.e., space defined and controlled by an authoritarian dominant order). He likens the way poiesis erupts under these circumstances to poaching: "Everyday life invents itself by *poaching* in countless ways on the property of others" (de Certeau, 1984, p.xii). Imposed "ways of using" products might reflect the aesthetic of the dominating order but does not necessarily signify dominated consumers, who may instead use their own agency to escape the dominating order without ever leaving it: "users make innumerable and infinitesimal transformations of and within the dominant cultural economy in order to adapt it to their own interests and their own rules" (de Certeau, 1984, pp.xiii–xiv). Life happens in individually unique and bizarre trajectories but can be graphed in terms of tactics and strategies.

De Certeau defines a "strategy" as "the calculus of force-relationships which becomes possible when a subject of will and power (a proprietor, an enterprise, a city, a scientific institution) can be isolated from an 'environment'" (de Certeau, 1984, p.xix). A strategy is a "victory of space over time," a rationally constructed spatialized institution that stands distinct in relation to its exterior. In contrast, he defines a "tactic" as "a calculus which cannot count on...spatial or institutional localization, nor thus on a border-line distinguishing the other as a visible totality" (de Certeau, 1984, p.xix). A tactic abides and whenever possible seizes opportunity, turning alien forces towards unintended ends. He suggests that this is the nature of many everyday practices. De Certeau directs our attention to "the uses of space" (de Certeau, 1984, p.xxii), which in the context of global media questions how, in an everyday life of electronic and computerized spatialization, people subvert the dominating spatial logic to assert and perform their own culturally informed counter-scripts. Through mobilization of communication, a "worker" may suddenly become a socializer, sitting back in his/her chair to relax and chat for a few minutes, or even appearing to be at work while chatting online. In fact, the production of data through chatting paradoxically turns leisure into work.

"Being" Canadian: Hide the Cell Phone (Scene 4)

I cram onto the southbound bus at Ossington Station. The motion of the bus jostles us slightly as we pull out. As usual, people are generally broadcasting rather disinterested body language. I'm sitting along one side near the back. Amid the random conversations taking place, I hear a cell phone ring.

"Hello?...Oh...Yeah, look...Oh, I see...Um hmmm...Well, I'm on the bus...I can't really talk right now...I'll call you in a little while OK?"

On the bus? Is that really a reason not to talk on the phone in Toronto? Why? There are no official rules against it, but come to think of it, it is something that rarely happens. In fact, I think I've said something to that effect myself, many times, and even in Champaign-Urbana. The bus continues on its journey down to Queen Street, sans phone conversations.

* * * * *

McLuhan described electronic technology as extending the human sensorium to embrace the globe. He claimed that this technological feat eradicates both time and space. He said those thus immersed become involved with "the whole of mankind and to incorporate the whole of mankind in us" (McLuhan, 1995, p.150), in effect participating intimately with each other in every moment. He used the term "global village" to describe electronic media as facilitating the involvement of people with one another around the globe. His famous phrase "The Medium Is the Message" (McLuhan, 1995, p.151) reflects this technocratic determinist argument, in which technology brings about social change by introducing new patterns and paces of activity.

Although he seems to have gotten some things right, history has not been kind in its refutation of many of McLuhan's predictions. Be this as it may, his idea of involvement remains a useful way of characterizing the nature of particular media devices, although with the effects of media temperature considered on a much more restricted scale. Involvement is a useful tool in considering the performance a given device demands of its user(s) in relation to another device. This notion becomes more interesting when looked at as exemplifying particular types of symbolic interaction. In particular, the notion of seduction can be invoked, which can be understood to signify simply how a technological device as actant calls to a user to form an alliance. The laptop seduces the user through the nature of how one engages with the machine, through the involved nature of typing and/or clicking a mouse while looking and possibly hearing. Communications technology seduces its users through the promise of human intimacy. The promise of mobile technology is the promise of human relationship.

"Being" Canadian: Hide the Cell Phone (Scene 5)

"Why don't you ever answer your phone?" I say into my friend's answering machine. "This is crazy!! ¿Donde estas? Ya estoy aqui… te espero en frente del restaurante… bueno... te veo prontisimo ¿sí?" (Where are you? I'm already here, waiting in front of the restaurant. OK, see you very soon, right?)

About ten minutes later, I see the familiar form of Luz, my friend from undergraduate studies. "¿Por qué no me contestaste?" I implore (Why didn't you answer my call?).

"Oh...¿Me llamaste?" she asks. "I didn't leave my phone on...I was in the subway...," she explains, as if it will be obvious that everyone in the subway should turn their phone off.

"Hmmm. Muy interesante...," I state. "A tí no te gusta usar tú cellular entonces..." (So you don't like using your cell phone then...).

"Aaah, well, Grant, I tell you, I really just don't care about the cell phone that much. You know...it's just a phone." Luz is never one to mince words. I laugh at her candor, like I always have.

"Yeah, it doesn't matter that much to me either, but it seems like people in Canada, or at least Toronto, don't like using cell phones very much, even though everyone seems to pay for one!" I explain with a laugh.

"Yeah, well...I just spent all day working my ass off, tengo hambre (I'm hungry)...let's go in here...," she says, pushing the door open.

* * * * *

As Innis pointed out, communications technology is concerned primarily with binding people together. Perceptions of both time and space are always somehow altered by mediation, and rather intensely when in translation by contemporary technologies. In particular, space becomes "dedistanced" (Heidegger, 1977), while time becomes "telepresent" (Virilio, 2000). That is to say, no time or place in the universe is considered inaccessible to a networked user in the present. Rather, through digital mediation, a user experiences omnipresence in time and space, able to as easily witness the birth of the universe as the death of our sun. Taken to its philosophical extreme, even conceptualizing a present moment becomes difficult to sustain in a world always contingent on situational multiplicity. Foucault's notion of heterotopia offers a way to better understand how multiple, paradoxical times and spaces can coexist.

Foucault explains heterotopias as "singular spaces to be found in some given social spaces whose functions are different or even the opposite of others" (Foucault, 1984, p.252). Heterotopia is thus defined by relationality as an expression of extension, which can be understood from McLuhan's explanation of media being extensions of the human sensorium. Foucault posits that governmentality is attained through sets of inviolable oppositions constructed and enacted spatially, always informed by a notion of sacredness (perhaps similar to Carey's argument about the performance of sacredness in everyday ritual). So for Foucault, space is heterogeneous, not a void, and we live inside a set of relationships delineating irreducible sites. Some sites can be defined by identifying the relationships that constitute them. However, there are two types of sites that stand in relation to all other sites, but in such a way as to contradict all other sites: utopias ("unreal space" with no "real" place), and heterotopias (a virtual space of mixed experiences behind surfaces, identified through mediated gaze). Foucault gives an example of what he means:

> [Heterotopia] makes this place that I occupy at the moment when I look at myself in the glass at once absolutely real, connected with all the space that surrounds it, and absolutely unreal, since in order to be perceived it has to pass through this virtual point which is over there. (Foucault, 1967)

Foucault elaborates six principles of heterotopia. The first principle is universality, which describes two types of heterotopias that ultimately encompass all aspects of human behavior. The second principle of heterotopia states that society can change the function of them (i.e., "relocate" them elsewhere or make them into something else), and the third principle tells us that heterotopias juxtapose several spaces that are incompatible in a "real" space (e.g., a garden in which plants incompatible with one another—that is, not growing in the same habitat in nature—are made to grow together). The fourth principle describes heterotopias as linked to heterochronies, specific time frames (i.e., they are temporally grounded). Fifth is the claim that heterotopias are exclusive, that they have doors, with either forced entry (compelled entry) or rites of passage, and they are demarcated as isolated places. Hence, one is always a guest in a heterotopia. Finally, the sixth principle tells us heterotopias "function in relation to all the space that remains" (Foucault, 1967), either as illusion exposing "real" space or as a "perfect" real space that stands in relation to the messiness of the rest of the "real" world. Mobile communication technologies bring with them the potential to turn any utopia into heterotopia. Whether we let it do so or not is a matter of choice.

"Being" Canadian: Hide the Cell Phone (Scene 6)

I'm really bored. After a great meal with good friends, I'm waiting for the subway at Pape Station in the Danforth area. There is hardly anyone down on the platforms. I look up the tunnel to see if I can spot the train's headlight. Because of the ever so slight curve in the tunnel, I can see clearly up to the next platform and beyond. I peer harder. Nothing. I step back to the wall, then pull out my phone, and take a seat on the standard ugly red plastic bench. I begin playing Bejeweled. Bejeweled is the game I play when I don't know how long I have, as opposed to Tetris, which I play only when I know I have to wait at least twenty minutes. I'm quickly absorbed into the little screen.

A few more people dribble onto the platforms. When I look up to meet their eyes, they quickly look away. They all seem to choose to stand in an assumed "bored" position. Finally, a few minutes later, the subway train arrives. I keep my phone open, enter the train, and take a seat. Continuing to play, I notice what is now a predictable pattern of surveillance from the other passengers. By the time I get to the Bloor/Yonge station, the train is too full, and I'm too surveilled to continue comfortably playing my little cell phone game. I put it away in my pocket, out of site, and assume a "bored" slouch in my seat for the remaining four stops of my ride.

* * * * *

With Foucault's criteria in mind, mobile technologies mediate time and space in several identifiable ways. For example, the cell phone "privatizes public spaces" (Puro, 2002, p.23), causing users to withdraw from public performance and/or from revealing private details in public. Thus, paradoxically, the cell phone also publicizes private space or at least private interactions. Meanwhile, a laptop can transform practically any space into an aesthetic symbolization of production, desacralizing even a church or mosque. Rules and norms regarding appropriate spaces and times of use abound, perhaps most easily exemplified by movie theaters and performing arts venues.

On the whole, mobile technologies work together to create and maintain heterotopia in at least five distinct ways:

1. by creating virtual social space (e.g., the laptop as cultural space and the cell phone as temporal teleportation of interpersonal relations)
2. by allowing imagined utopias to be located within the messiness of the actual space one is in
3. by distancing one from their own location and de-distancing others
4. by re-spatializing/rearranging architectural space
5. by communicating existing social codes (e.g., conveying prestige and status, and as actants when talking [an act of translation] about technologies)

This fifth characteristic most clearly relates to the emphasis Carey put on cultural practices. By stressing existing social codes, heterotopic experiences have the potential to be defined and judged morally, per the social context in which they take place.

"Being" Canadian: Hide the Cell Phone (Scene 7)

It's Saturday evening. I'm at a restaurant to have dinner with the Baldizón family: Christian, Reinhardt, José and his wife Veronica. We're having a band reunion of sorts, minus Ryan, our former singer. It's always fun to spend time with them; I always feel like I belong with this crew. While we're waiting for our meals to be served, Veronica asks what I've been doing, how my research is coming along. I explain the interesting Canadian phenomenon of "hiding the cell phone."

"Hmmm, that's really interesting...," she says with the smile of someone laughing at oneself, "I hadn't really noticed it before, but now that you mention it, I guess it seems right..."

"Yeah, it's really strange in a way...," I say.

"Well, I DO have one myself," she confesses, "but I usually keep it in my purse and don't like to have it on too much..."

"What about the rest of you?" I ask in general.

"Yeah, I got one right here...," Christian answers, patting his chest pocket. "Yeah, I don't know if I try NOT to use it, but I usually only use it for work stuff...I don't use it much for personal calls," he qualifies himself.

"Reinhardt, what about you?" I ask.

"Yeah, I got the cell phone, you know...," he says, almost embarrassed, "but really, I just carry it around, I try not to use it..."

"Oh, why not use it?" I ask.

"You see! He's doing his research right now!" José says to Veronica. I look at him and laugh in agreement.

"Well, maybe it's because it costs money...," Reinhardt answers, "but I don't know...That doesn't make sense because I have a monthly plan... Uh, I don't know!"

"José, where's your phone, man?" I ask.

"Right here man," he slaps his thigh pocket. "And it is OFF right now!"

"Off? How come?" I ask.

"I don't know, man," he says with finality.

I explain the various theories I've come up with to try to understand this behavioral pattern and recount my conversation with Walter. The three brothers mumble a restatement of their previous comments, but Veronica looks at me thoughtfully, smiles, and says, "Well, Grant, honestly, I think I just feel it is rude to talk on the phone in public...You know? It doesn't seem polite to have a private conversation in a public place. I don't know, maybe I'm just too Canadian, but I guess that's why I don't like using the cell phone very much." We four men sit in stunned silence for a moment, thinking, as if in a soundless oasis listening to a clock ticking, surrounded by the din of the restaurant's crowded chaos.

I feel a lightbulb go off in my head. Veronica has just explained my own strange penchant for seeking out private little corners and nooks when I answer my cell phone. The brothers in a sonic jumble voice their agreement with her explanation, as do I. As the food arrives, we effortlessly move on from the topic, enjoying each other's company, gossiping, and recounting our past glories.

"So Grant, how are things going in the States?" José asks. Without further pause, we move on to other concerns.

"It's good...things are going well...actually, I'm getting ready to go to Asia now...," I answer. "You guys should think about coming over there too!"

CHAPTER FOUR

Ontology, Technological Mobility, and "Belonging"

Revealing and Epiphany (Scene 1)

It is winter 2003. I am living in transience; temporarily abiding with my brother in the apartment we used to share, on break from the University of Illinois. It has been years since I "officially" moved out, but he is my magnet of home, so it is to him that I return. His roommates are gone for the holidays. I am sick. It is my second round of the flu this season. Just two weeks earlier I was quite sick in Champaign-Urbana, so bad that I sat up with a start in the middle of a feverish night, all strength sapped from my body, realizing I could actually die from the illness. In what I imagine to be a perennial question for all people who live alone, I wondered how long it would take for someone to notice I was missing and to find my body.

It's now Sunday night, a couple of hours after dinner. I suddenly find myself getting violently ill again. I go upstairs and lay down. I've been sleeping up in Joe's room on the top floor of the three-story building, which used to be my brother's back when I lived there. Now it is me and my laptop's turn to occupy the space for a few weeks. Joe is a professional jazz drummer, and his spartan room reflects his own transient lifestyle: a futon mattress on the floor, a dresser with a couple of drawers hanging open, some drumming hardware strewn about the floor, and a couple of suits in his closet. My suitcases seem right at home in the thin clutter. My laptop is the only thing that keeps giving me some sense of self, spatializing at least a few of the icons of my own sense of being.

There is no problem finding random wireless Internet signals in this busy part of Toronto. At any given moment, I can usually find two or more signals to tap into. I can move from room to room in the apartment and maintain connectivity. But up in Joe's room there are even more options. Who am I trying to connect with? Of course, I have my brother, who nurses me as best he can. But for several days, in between hot flashes and chills, in between stints downstairs in the living room and the kitchen and extended trips to the bathroom, I keep my chat line with friends as busy as I can, check and write email, and when I have the strength I read the newspapers online. All things considered, I'm doing fairly well in my social life for someone who feels deathly ill. And then, tragedy strikes. The power adapter quits working. It is a disaster of unmitigated proportion. My battery has about three hours of charge in it. I am suddenly alone, with the cold winter wind hovering above Bloor Street. What will I do now? I look ruefully at the pile of communications philosophy books I've brought along and see my plan to finish the philosophy of technology essay I owe Prof. Christians mentally dissolve.

* * * * *

Heidegger tells us that technology is instrumental by definition, but that this is different from its essence. Even so, understanding the instrumentality of technology is an important part of understanding its essence as it relates to human "being." He suggests that the instrumental aspect of technology is the confluence of four very practical modes of setting causality into motion, or "occasioning," which contribute to a fifth overarching instrumental function he calls "poiesis" (p.10). Human agency comes into play in this causality as the driving force of poiesis. Poiesis, defined with the gerund phrase "bringing forth" (p.10), is every occasion that something makes itself present or is made present, described succinctly with another gerund as presencing. Poiesis is grounded in "revealing" (p.12), meaning technology brings forth and reveals truth; technology allows truth to present itself. Understanding truth is yet another formidable challenge, the nature of which Heidegger describes as ambiguity, and seeing danger.

In contrast with the mechanics of technology, Heidegger states that human thinking is reactionary and responsive to things presencing. People are ordered to act by the presencing of things with already existing conceptual paradigms that contain their own imperatives. People are ordered to drive technology forward with their own human agency. Humanity is in effect the engine of technology, enframing (i.e., constructing and maintaining a conceptual paradigm of) the universe that presents itself to us. He claims that because we rely on always already existing conceptual paradigms to make sense of the world as we encounter it, our relationship with technology seems to naturally turn into an ostensibly self-contained and self-perpetuating cycle. Technology is thus a potential trap, in that participation in this technological enframing paradox without questioning or challenging it obscures the fact that it is our own agency in the form of poiesis that drives technology onwards. However, the opportunity to discover the truth of being is found in the contradiction of enframing what presences or unconceals itself.

Revealing and Epiphany (Scene 2)

The next morning, I take the adapter to the Apple store. I find it is covered by warranty, but it will take several days before I can get the new one. It actually takes a week. Meanwhile, I decide to read all the Heidegger I can get my hands on. In my feverish daze, I scour the city for his writings. I voraciously read everything I can claim for myself. I think long and hard on it. I meditate on it. I write notes and rewrite them, and I reorganize my thinking. Then, one night, in the middle of a fever, I rather gently find I understand how everything fits together. No bells and whistles, no explosions, no gesticulations or prostrations...just quiet realization, alone, in my sick bed. And not just Heidegger, but everything: my years of questing after spiritual knowledge, my intrigue with time and space, my fascination with society, my daily rituals, yoga, Vedanta, fasts, my enjoyment of philosophy and theory, my physiology, the relationships I've had in the course of my life, my family, the food I eat, war, peace, love, joy...I realize that I just know it now. I feel the illusion of my existence, and how easily my being can dissolve—it is, in fact, constantly dissolving, just as my writing plans had earlier fallen to pieces. With me will die all those memories and knowledge, yet living, all is simply illusion anyway. But I also feel the immortal kernel of life energy around which all else is conglomerated. I understand how my own mind seeks to make sense of it all, to construct fantastic castles in the sky that impose order on otherwise chaotic and random phenomena. I reflect, "It is the enframing that one holds onto." At the same time, I now see that none of it matters at all. Laying there facing the absurdity of life and death in the middle of the night, I realize I do not "belong" to the world, yet the world is all that I can ever know until I leave it.

Heidegger already said it all, as perfectly as can be (which is to say imperfectly), but then, so did all the others who helped me get to this place, all the philosophers and holy teachers, and even my little laptop computer that, if it hadn't broken down, might have kept me on track to read enough Heidegger to come to this realization. In fact, I apprehend, that all the questions I had have been answered. I have learned everything I set out to. I realize the arrogance of the proposition but laugh out loud because it's quite simply true. Is this how a champion feels? I realize there is nothing left for me to do, no theory or philosophy left for me to "figure out" or reveal to the world. I find myself suddenly relieved of the burden of my questions. There is nothing left for me to write, except what I'm told I must to qualify as someone who "knows" something—a test of my ability to enframe, perhaps even to be the engine of poeisis. I realize I cannot talk about even this experience, since most will not understand and will dismiss me as arrogant or foolish or both—and those who would understand, well, what difference does it make to them? To talk about it is only to pitifully attempt to hold onto some shred of an imagined ego that I now know never existed in the first place.

A couple of days later at lunch in a Vietnamese restaurant in Chinatown, I look at Nitya across our bowls of noodles. It is a bitterly freezing cold January day. He is my eternal friend, my spiritual teacher and mentor. He is telling me about something that was important at the time, but which now escapes me. His daughter? His yoga school? His general state of being? Of course, he would say, "They're all equally important! And unimportant too!" I look at him and listen, silently mulling over what I want him to confirm.

Questioningly, I state my realizations:

> Since all of this life is illusion, the only thing one can know is illusion; since all experience and knowledge in this world is illusion, oneself is always an absurdity and an illusion anyway; the spiritual being simply "is" and, whether we describe it or not, is futile since all description is ultimately inadequate; all that matters then is physiology and psychology in the moment we are presently in, to help and allow oneself and others to feel as good as possible in every instant. And why? Quite simply because that's the kind of world I want to live in, just because that aesthetic pleases me, no other reason.

He looks at me with a gentle smile on his lips, stares seriously into my eyes, and says quietly, "Yes, that's right."

I look at his chapped lips, start laughing to myself, and can't quit giggling like a kid. It's funny that I should need confirmation of this, and I realize I was really just looking to feel less alone in my being able to accept this. This man is my teacher and friend, my guru, my brother, someone I love as any disciple should love their master, and he has just affirmed my realization that my years of questing for spiritual "knowledge" have been only about satisfying my own egoistic desire to "know" something that other people don't, and that my own "beliefs" are just stories I tell myself to keep my psyche optimistic and my actions purposeful. In those three words, he has confirmed my enlightenment for me. I'm overtaken with giddiness. NOTHING matters, so I just need to concentrate on constructing myself and the world I live in to reflect as much of what I want to see and experience as possible.

Nitya talks about how the traffic can be affected by something as simple as a movement by him or me at the table where we are sitting. I find I no longer care at all about acquiring that kind of "mystical" power, that all I care about is the fact that he is talking to me, sharing this moment. I don't know if he can tell what has just happened to me, but it doesn't matter anyway. It is MY moment to experience the temporal paradox of eternity, to know myself and everyone as Godlike. We pay the bill and walk south down Spadina Avenue. I part ways with him to catch the streetcar home. I watch him walk away from me, towards his yoga studio around the corner. I realize that although I will certainly continue to practice with him, I've just had my last "lesson." My giddiness renews.

Figure 4-1. Swami Nitya Muktananda in His Yoga Studio.

* * * * *

The epiphany of enlightenment is to turn one's vision outward, to both see the starkness and apprehend the significance of oneself as part of an infinite, immortal, overwhelmingly unenframable universe (Heidegger, 1977). Heidegger explains that modern technology is a particular schema of revealing/enframing that is in its nature an ordering that treats the natural world as a "standing reserve," a stockpile of goods perpetually "on call" for human uses (pp.15–17). The standing reserve is a theoretical setting apart of "things" into a conceptually closed system of ordering (i.e., enframing). Theory uses the concept of nature to set things apart into the standing reserve, treating all present or potentially present things as enframable, within the realm of human comprehension and classification. Things in the standing reserve dwell in a status of being perpetually on call, at the behest of human use. Although it is the essence of technology, enframing is itself not technological. However, by revealing humanity itself is also ordered as part of the standing reserve, enframing is the challenge that motivates technological activity; it is "the real" revealing itself as "standing reserve" (p.23). Humans are made to dwell in the standing reserve right along with the entire "natural" universe.

Heidegger claims that history is written as destiny, an "unalterable course" (p.25) that objectifies time to make it accessible to be worked upon. Technology is historicized as fatal—the unalterable course of humanity, trapping us with the enframing that we as people only have "use value" in the service of maintaining and creating technology and have no other logical reason to exist. Freedom for humanity is consequently not about causation or will, but rather about being in the realm of destining in such a way that one is able to experience our technologically enframed existence without being "constrained" to comply with the seemingly natural order (p.25). Blind obedience to our already existing conceptual paradigms in the guise of what is natural, in fact, obscures the truth of our existence and is, therefore, dangerous, creating the self-maintaining trap explained above: "All mere organizing of the world conceived and represented historiographically in terms of universality remains truthless and without foundation" (Heidegger, 1977, p. 48). Freedom is to be found in the dressing up (enframing) of truth in such a way that the truth shines through and lets the enframing be seen as enframing. Any other way of concealing (i.e., effectively letting the truth remain obscured) is dangerous.

In this endless recycling of the same paradigmatic enframing, God (that which is unexplainable) is defined reductively as a foundational cause (potentially *any* unexplainable cause), and humanity remains simply trapped in the conceptual standing reserve. The unexplainable (God) thus conceptually tucked away from interfering with theoretical activity (just as the refusal to define freedom accomplishes for many theorists of technology), humans come to believe themselves the only source of agency, enabling a solipsistic fantasy that humans have created and maintain the entire universe.[1] But here the paradox of enframing presents an opportunity. Although technology is the fate of humanity and enframing tempts us willingly into its trap, freedom can be gained only by employing technology.

Revealing and Epiphany (Scene 3)

I go to the Apple store just down the street and pick up my replacement power adapter. When I return home, I plug in my computer and immediately go through all my regular rituals of checking email and reading the news online. I look for friends to chat with. Lying on the bed, I look at the pile of books and readings waiting for me to shape them into an essay, and I realize I have to rethink my entire education. I cannot betray Heidegger, for that would be a betrayal of my own essential understanding of everything. But then…there is nothing left for me to say anyway! I decide that an entirely new approach is called for, but for the moment I can't imagine what. At the very least, I know I will need to summarize Heidegger's writings, and that in itself is no small project. This becomes my primary task. I abandon the tall column of paper for a half-dozen books and begin the arduous task of paraphrasing a master.

I go out and work in coffee shops down the street. I work in libraries. I work at the kitchen table. I work in the morning, afternoon, evening, and night. I carry my laptop with me everywhere. I become routinized to work in public spaces, acclimatized to busy places—sometimes surrounded by others busy on their own laptops, sometimes not. I work in Toronto, and I work in Illinois. I assemble a lengthy set of notes on Heidegger and work through my conundrum, deciding that, knowing what I now know, the only reasonable way to proceed is to take Heidegger's challenge seriously.

1 Which, as Lunenfeld points out, gives us the extreme dystopic/utopic positions in which we either are being destroyed by the technologies we create or are creating our own emancipation through those same technologies.

"Poeisis…," I repeat the word in my mind like a mantra. I decide all that is left is to find a way to elaborate in text the eruptions of truth in everyday life, to identify and describe the moments in which the enframing is revealed for what it is, when pure being stands to work on its own. "Poeisis…look for those magic moments…"

* * * * *

Tending to ontology requires that the language used to enframe Being enable thinking of a nature in a way that a solution is possible. Of course, any enframing is the turning of thought inwards, presenting the danger of a solipsistic slide in which enframing is mistaken for Being, and the truth of Being thus becomes obliviated in the game of designing ever more sophisticated enframings: "all mere willing and doing in the mode of ordering steadfastly persists in injurious neglect" (Heidegger, 1977, p. 48). This doesn't mean there is no truth, but rather that such oblivion serves as a "safekeeping" of Being. Obliviously, one might exist in the world without active awareness of the truth of Being. One might not even want to learn the truth about an objectified "being" that seems to abide outside oneself, and which distracts attention from attending to the intimate truth of one's own Being. But then, why should one care about this at all? Quite simply, because there are questions and fears such as those outlined by the aforementioned dystopians and utopians and others—there is a language that enframes the issue—and there are consequently some answers. Enlightenment *is* possible when the right phenomenological tools are employed, and it is the job of an enlightened humanity to attend to the truth of Being: "Only when man, as the shepherd of Being, attends upon the truth of Being can he expect an arrival of a destining of Being and not sink to the level of a mere wanting to know" (Heidegger, 1977, p. 42). Bringing about understanding of Being's truth is the aim, by challenging the mode of thinking such that Truth makes itself visible. All the while, being is neither in danger nor dangerous, not touched nor worked upon: "Being…is not brought about by anything else nor does it itself bring anything about" (p.44).

Revealing and Epiphany (Scene 4)

I stand at the kitchen window with a cup of coffee in my hand, looking out over the roof below the apartment. Soft white snow has piled up on the green steel fire escape stairs that descend over the window from the floor above. The bright morning sun blinks off the snow, making the day feel brand new. My laptop sits gaping open on the table beside me, broadcasting the *Toronto Star.* "Hey! How you doing?" my brother asks, walking into the room behind me.

"Oh, pretty good...," I say, sitting down. He pours himself a cup and takes a seat across the table from me, sharing this moment of my morning ritual.

"Is it working OK now?" he asks, referring to my computer.

"Yeah, it's all good...," I tell him, "but man, I got a lot to write..."

"Good...good...Well, I guess that's what you asked for, eh?" he ribs me.

"Yep, that seems to be what I signed on for...," I acquiesce, taking a sip of coffee. "Gotta be careful what you ask for, 'cause sometimes that's exactly what you get, eh!" I say with a big smile. The cold sun suddenly flashes off the snow, lighting the room in a happy glow.

* * * * *

Heidegger (1977) describes that freedom in relation to technology is possible only by understanding that the essence of technology is a crucial and inseparable element of our human existence. Human freedom is not to be found in rejecting technology or in thinking of technology as something outside of human "being," but rather in understanding how the essence of technology is a very important facet of ontology. Hence, to understand human freedom requires an understanding of the essence of technology. Truth can be found only through technology, since truth itself is ideal and needs to be demonstrated, brought forth, and made to present itself. Self-inquiry, intellectual methodologies, ceremony, and ritual are ways of bringing forth the art of the self, the poiesis of being a person. To inspire such causality, people must first be in a spatial position that attends to their relationship to "Being."

In my interpretation, once a "rift" has been conceptually established between humans and technology, or between one another as humans, one's ontology ironically becomes both in need of "repair" (or Heidegger might say it needs to be attended to), and at the same time un-repairable (since there has been an allocation to the oblivion of the standing reserve). The process of creating ontological security—keeping things apart from oneself—enframes the world in such a way that one hides the truth from oneself. In such a configuration, can there ever be a true sense of "belonging"? When one's life experience consists of constantly jumping from place to place, with no permanent house called home, how does one make sense of others' apparent need for them to "belong"?

Durkheim (1996) suggests the Western discourse of religion is concerned with defining and demarcating the spaces of the sacred and the profane: "the sacred and the profane have always and everywhere been conceived by the human mind as two distinct classes, as two worlds between which there is nothing in common" (p.91). Excusing his claim to all of time and space on behalf of European theorization, there is a very basic implication of this categorical division that influences spatial organization: "Sacred things are those which the interdictions protect and isolate; profane things, those to which these interdictions are applied and which must remain at a distance from the first" (p.92). Any moment of contact in which the profane "other" breaks what are taken for granted to be the rules of communication shakes the very foundation of the West's dominant discourse of civilization and breaks the dominant mode of enframing. Such a redefinition of the rules of discourse, such a disruption in the dominant enframing of the world, is extremely uncomfortable for the Western subject—a moment of terror, threatening to expose the silent assumptions of the system to one who acts obliviously as the engine of it, but who remains ignorant of the possibility of changing it. Although overly simplified, this interpretation of Durkheim I offer as a schematic of the dominant modernist ideology informing the Western structuration (Giddens, 1986) of the world.

Reflecting a Cartesian body/mind dualism, the discourse of the sacred and profane serves as a foundational cultural blueprint for both the physical organization of space and the organization of the orthodoxy—or "right-thinking"—of modern Western society. Holding apart danger, keeping it untouchable and, therefore, unconfrontable and unmanageable, results in the seeking of safety through everyday routine (Giddens, 1990). Predictability and self-possession are the hallmarks of everyday life. Anxiety is held at bay through mundane personal practices, building a trustworthy and reliable world through predictability. Routinization of social conventions—ritual, in this sense—is motivated by the construction and maintenance of security. Prolonged "ontological insecurity" (Giddens, 1986, p. 62) with the constraint of strictly imposed obedience is devastating to one's sense of self, but Heidegger is at pains to explain that such constraint is antithetical to the truth, as it seeks to substitute one order for another.

A Letter to a Friend

It seems to me like I've never *really* belonged anywhere, or stated differently, that there's nowhere I don't belong. As well as I can remember, this transience has always been a part of how I perform myself, and I don't want to forfeit this habit and the freedom it gives me. I've moved around so much that I've had many opportunities to transform and play with my identity—when I was a kid, I used to even imagine how I was going to "be" in the next place I would live in and plan how to enact that newly imagined "me." Others who moved around a lot in their childhood have related similar stories about themselves to me. Psychologizing myself, I would hazard a guess that this is part of a coping strategy, to deal with the uncertainty of never having a secured or permanent space called "home." Perhaps it's somewhat predictable in that I was often thrust into "communities" that I didn't feel were willing to accept me anyway—to preemptively refuse them first and mask the pain of having been torn from my former lives—but this is a digression from the main topic. My main point is that, for whatever reasons, the effect is that I'm resisting the term "belonging" in relation to myself. I suppose this is the ethical or epistemological assumption that refuses to define the broad contours of the interpretive community I belong to. It simply isn't a concept that has been supported by my lived experience, and as such it is not part of the world I imagine myself to inhabit.

Though I'm a "city" person now, I was born and raised during the most formative years of my life in a land without fences. I was taught by example that I could go wherever I wanted to go, be whoever I wanted to be, use whatever the earth offered up to me, and think whatever I wanted to think. These are four lessons I have chosen to carry forward with me as I have progressed through my process of healing from all those other things I've chosen to leave behind. I am a spiritual person. Ceremony and prayer are important to me. When I pray, I talk to my ancestors (those who came before and will come after), and I talk to my helpers. I pray for my family, my friends, my teachers and students, and myself. I pray to be gentle, generous, wise, and strong. I pray for peace in my heart, and I pray to spread that peace in all the directions. I pray that my actions and thoughts be guided by the wisdom of my ancestors, and that I affect the world during this imagined existence in a positive way for all life. Do the spirits of the dead and the yet to be born constitute an interpretive community? Does an ontology of imagined existence allow for "definition" and "framework"? I suppose the obvious answer would be yes, if one imagines it so, but as I said, I come from a land without fences; that's just not how I "know" the world.

* * * * *

Denzin (2001) defines two categories of understanding: spurious understanding versus true or authentic understanding. "Spurious understanding" is "when an individual only superficially enters into the experiences of another" (p.139), mistakenly projecting his or her own preconceptions onto another person, and thereby failing to actually grasp the other's point of view. This sounds very much like an uncritical reproduction of Heidegger's description of enframing. "True or authentic emotional understanding" happens "when one person enters into the experiences of another and reproduces or experiences feelings similar to those felt by the other" (p.139). One dwells in the experience of the other, the experience is truly shared, and authentic understanding is thus possible. This emphasis on the relationship of experience with authenticity seems to be in keeping with Heidegger's description of truth. Although Denzin is primarily concerned with applying his definitions for the purpose of understanding interpersonal relationships, I would like to borrow his classifications to describe the experience of one's relationship with oneself.

Heidegger (1996) describes the authentic self as "the self which has explicitly grasped itself" (p.121), which is not the self of the everyday, but which paradoxically can come to grasp itself only by dwelling in the everyday. It is the self that has fully experienced the paradox of being: enframed and dwelling, but not constrained to obey. This is the self that can be understood by being enframed in such a way that it stands in contrast with the "spurious" self, which has enframed the essence of its own being and relegated it to the oblivion of the standing reserve. The spurious self embeds distance into any conceiving of an authentic self, paradoxically objectifying one's essence while at the same time preventing one from truly gathering the elements necessary to experience and know oneself authentically. One becomes buried in the "normal" structure of everyday life. In terms of subjectivity, the danger of mobile technology is thus that by embedding distance in a hyper-mediated everyday experience, it conceptually reduces the chances of experiencing the universe on its own terms as an active entity, preventing the potential of gathering the universe to oneself in order to authentically understand oneself. The mobile subject requires self-objectifying self-surveillance, self-directed vigilance in maintaining the dominant symbolic order. Foucault's panopticism is extended by this technologically enhanced self-surveillance, and thus mobile subjects including myself participate in their own objectification even as they appear to escape and challenge institutional apparatuses. Escape itself can be only from one network configuration into another, more stable-feeling set of associations.

CHAPTER FIVE

Mobility, Globalization, and Culture

Globalization: "the intensified and accelerated movement of people, images, ideas, technologies, and economic and cultural capital across national boundaries."
— Cameron McCarthy et al.

Technography and Belonging in Vancouver (Scene 1)

Vancouver is a lovely city in the fall. The rich flora acts as a living backdrop of vibrant colors, leaves in various stages of changing from green to yellow to red; in places, you might even spot a street hockey game or two. I'm here to present a painstakingly prepared paper (Kien, 2004b) on Heidegger's philosophy of technology at a conference put on by the Vancouver Art Gallery (VAG) as part of the city's New Forms Festival (an early draft of the appendix). The conference is called "Technography," a seemingly perfect coincidence for the direction I've decided to take in my own research, and as natural a choice of theme for a Canadian conference as one could ever imagine. The conference is held in the upper levels, in which the remnants of the courthouse that the VAG once was remain. The main gallery downstairs exhibits Bruce Ma's futurist installation "Massive Change," which after touring I describe to myself as an exposé of "nano-colonialism." The hotel I found on Priceline is coincidentally right across the street from the gallery. I don't spend much time there, but I'm nonetheless disappointed there is no wireless Internet in the room. I don't even think to look for something so primitive as a wire, so it takes two days before I realize there is a live Ethernet cable tucked beside the desk in my room. No problem, I've been getting my CBC dosage from the television itself rather than online, and the conference is taking up the morning time when I would normally read my newspapers online. In the afternoon and evening, I have strolled through the vicinity of my hotel, unsuccessfully looking for a wireless coffee shop. I barely see anyone using even a laptop, never mind the Internet. Cell phone habits seem to be typically Canadian: people give the impression of carrying them but keep their use to a polite minimum. I don't use my own phone, which is internationally roaming in this situation, but I keep it with me and have it switched on anyway in case someone has an emergency. I note the strange technologically induced disjuncture—to be internationally roaming in my own country. In contrast with the wirelessly barren cityscape, the conference rooms across the street are fully wireless.

Between hobnobbing in the reception area and attending sessions, I notice a real absence of cell phones during the conference, even though there are numerous papers discussing the implications of these technologies. Anywhere else I would expect the hallways to be filled with busy cell phone conversations between panels, but not here. I take advantage of the conference wireless as much as possible. It is now the final afternoon and I'm sitting in the main conference room—a converted courtroom—checking email and chatting on MSN. Machiko Kusahara has just finished her plenary presentation, mentioning how she witnessed Japanese girls using their cell phone cameras as periscopes in order to catch a glimpse from the crowd swarming around a famous singer. While I type a little note to myself about this, the final session is being set up. It is going to be a panel of native elders speaking about native culture and technology, or so the program says. The room slowly fills up around me as I check my email, and it is soon time for the panel to start. In the traditional, ritualized format, they will each take their uninterrupted turns talking, and everyone else will respectfully listen.

* * * * *

Global trade networks have helped along the enlightenment project by making capitalist flows more laissez-faire, influencing language, bringing about the emergence of communication through signs, and imposing standardization that permits communication to become more widespread (Mattelart, 2000). With this has come, among other things, an international division of labor, unified areas of trade, the creation of symbolic representations of the industrial "nation state," and a planetary clock. Global interdependence has evolved enormously, and there has been an "awakening of planetary consciousness" (Mattelart, 2000, p. 66), but this global consciousness has tended to be dominated by standardized information structures that seek to create a "free" global market. In practice, globalization is a contested spatialization on a world scale, with the potential for both positive and negative effects. Although it promotes "structured inequality" (Axtmann, 1997, p. 34), it also has the potential to influence progressive change in terms of global human rights and may sometimes even improve the economic circumstances for some (albeit normally a select few) in poorer economic regions of the world. Of course, often the opposite is also true as transnational corporations force indigenous populations off their traditional lands and out of their traditional lifestyles, creating the very impoverished demographics global economic development rescues to justify itself. Understandings of identity and nation in the circumstances of globalization are complicated by the dynamics of mass migration (forced and voluntary), movement of cultural and economic capital, and electronic (digital) mediation. Philosophical, theoretical, and methodological challenges arise in relation to "traditional" modernist notions of nation and citizenship. The contrast of identity and nation suggests a modernist inside/outside, micro/macro dualism: citizens inside the nation, non-citizens outside of the nation, identity as personal, nationalism as public. As McCarthy et al. (2003) express, feelings of belonging—uncertain or otherwise—arise within structures of domination that resist change and fluidity that characterize processes of personal identification.

Technography and Belonging in Vancouver (Scene 2)

As the session proceeds, the first elder talks about how the "bad" energy of computers adversely affects people, and the third one talks about how using computers is a waste of time, but for the most part they don't say much at all about technology. Rather (perhaps predictably, since I've been to many of these elders' talks in different contexts) the attention is mainly on rambling tails of hurt, injustice, and woe. With 500 years of colonial history to draw from, there is no shortage of pain to tell about. The final speaker, the only man on the panel, basically tells us that technology is a European invention that tears the soul from the body. He uses the automobile as an example, relating a story about how someone's spirit once got knocked out of the body in a car accident, requiring a ceremony at the accident site to send the spirit back to the body. I think about the historical fact that mass-manufacturing of automobiles is actually "native" to North America—in fact, the horse came from Europe, but he doesn't seem to have a problem with that particular vehicle. "Is it easier to form an alliance with a horse, or an automobile?" I wonder to myself. "Definitely an automobile," I decide, "horses have personality..." My thoughts are abruptly interrupted, as he hollers to the room in the rehearsed, pained/pleading mellow-histrionics of a victim looking for guilt-induced acquiescence: "You've left your spirits behind...GO BACK TO WHERE YOU BELONG!" I was always taught to respect elders, but as I sit with my hands on my keyboard, multitasking, absentmindedly dreaming and chatting on MSN in spite of the melodramatics going on in front of me, I decide it's pretty hard to respect disrespectful behavior. It's obvious where his pain is coming from, he's been quite forthcoming about that, but what's with this quasi-geneticism? Where is *it* from? For some reason, I don't feel it "belongs" here either. I sit and reflect that all the native spirituality I was taught always emphasized that every spirit in the universe is equal and that identity is arbitrary and political; has this man himself succumbed to some kind of fascist European propaganda that orders the universe according to some kind of blood origin myth? "This guy needs to read Gilroy's *Against Race*," I type to my friend in my IM browser.

"Go back to where you *belong*...," I turn the words over in my mind. "Belong...Where do I 'belong'...? Where does anyone really 'belong'?" I wonder to myself. "Man, xenophobia is just soooo stupid!" I type to my friend. The conference organizers...everyone really...act as if nothing unusual has happened. I realize it's because nothing unusual did happen, just that a change took place in me: a realization. I'm angry. "I'm tired of watching people perform their pain to absolute strangers and of being insulted if I don't want to listen," I type to my friend. "It's not just this guy, it's a lot of people and a lot of issues," I continue.

"Yeah, I hear you," comes the reply, "It's hard to be patient sometimes." Sure, people have their reasons and needs, but in that moment I realize I'm just tired of helping enable the litany of public therapy by being enlisted as a captive audience to it. I came here to discuss technology and, like several others, spent money out of my own pocket to make this event a "success." It just doesn't seem right to sit and be scolded, taken as an object just so someone else can be told they are "empowered."

"Don't worry, it'll be over soon, won't it?" comes the message. "When do you go home anyway?"

"You mean home Toronto or home Illinois?" I have to ask. As it happens, my friend is actually in New York, so I can't see much sense in the question anyway. "Illinois Monday, Toronto maybe November. U?" Mentally, I sigh, "Go back to where you belong...Shiiiiiit..."

A couple of audience members try to get the elders to talk more about the roles of technology in ritual and culture, and other such themes relevant to the conference. But the elders stick to the performance they know best. "I don't know anything about that stuff," one states bluntly, "I don't really know why I was invited here and what I was supposed to talk about today." As if signaling the end of the conference, my battery dies and I fold up my laptop. A few minutes later, the conference organizers, flanked by a Canadian government official and a big shot from the VAG, step forward and present each of the elders with expensive gift packages that include Bay blankets and other memorabilia. Any one of the gift packages could have given a homeless person shelter for a month, but ritualistic tradition takes precedence over such do-good pragmatics. "The ritual justifies their involvement...," I think, reflecting on James Carey's work. With everyone smiling at each other in self-congratulatory euphoria, the conference ends with this reenactment of an idealized thanksgiving. As I'm about to stand up and leave, I hear a cell phone ring. I watch from across the room with a little grin of ironic amusement as the final speaker pulls a clamshell flip phone out of his pocket and answers it.

* * * * *

Reflecting on the context of globalization, Appadurai states that "we are functioning in a world fundamentally characterized by objects in motion" (2001, p. 5). He explains that the contemporary manifestation of global capital employs "strategies of predatory mobility (across both time and space)" (Appadurai, 2001, p. 18). Such strategies compromise the abilities of people identified within a singular localization to comprehend, foresee, or oppose such predatory capitalist strategies: "globalization confronts local traditions throughout the world, influencing all levels of social life" (Cvetkovich and Kellner, 1997, p. 1).

Globalization is a grandiose *concept*, while the local is a personally *lived experience* in which ideological and cultural concepts inform behavior and justify individual interactive performances (aka cultural scripts). Analytically useful as these distinctions might be, it is impossible in practice to extract one from the other. Modern capitalism has always been a global venture, founded on the mass migration and forced relocation of millions, and on trading cultural and economic goods throughout the world. Even the present multinational systems of production have the British Empire's textile trade as a historical referent, and identification—racism in particular—has always been part of the structuration of these global enterprises. Digital electronic mediation and advances in transport introduce a new scale in terms of pace and distance, even a new product in the information economy, but the fundamental international business model remains intact. To use McLuhan's language, the physical systems have changed in terms of the scale, pace, and patterns of human affairs (McLuhan, 1964). Contending with the effects of these changes in the discourses of philosophy, theory, and methodology happens in the sphere of culture, in the ritual processes of making meaning in the way that James Carey (1989) describes. The physical and the metaphysical come together in and through practice, moving with the appearance of unity towards some elusive, unspecified, and always mobile destination. As Roy (2007) explains, physical, embodied interaction gives affirmation to what otherwise remains purely virtual community.

Wireless Guerrillas in the Jungle (Scene 1)

In 1994, indigenous people in the southern Mexican state of Chiapas shocked the world by forming an army and emerging from the jungle to take control of their globalized destiny, just as the Mexican government constitutionally abolished the right to collective land holdings and NAFTA came into effect. The media savvy Subcommandante Marcos quickly set about branding the image of the Zapatistas, becoming a willing posterboy for what became dubbed by *The New York Times* as the world's first "postmodern" war—a war of aesthetics fought through globally networked alliances. Marcos and the Zapatista council were not foolish, knowing full well that the only way to halt the overwhelming Mexican army was to aesthetically internationalize the confrontation. Through satellite hookups, Marcos was able to send out regular emails detailing the Zapatista's demands and make known on a global scale the conditions people endured in the Mayan communities. The image of a mounted man in a ski mask, pipe, and headset circulated freely around the World Wide Web.

The masked rebel became famous. Marcos's wireless laptop became a symbol of revolutionary empowerment, a means of equalizing voice in the global media system. Wireless in this context means everywhere, anywhere, any moment, every moment, even in the jungles of Chiapas. While he sent out regular reports and updates and manifestos from the jungle, Mexican authorities responded (in part) by attacking the signifiers, lamely demanding he take his mask off. He and the other EZLN soldiers refused, saying that if they take their masks off, they simply become invisible native faces, to be as unseen and ignored as they were before the uprising, as voiceless as the domestic servants populating the houses of the rich. Using this aesthetic strategy, the Zapatistas were very successful in gaining international support, and countless web sites, list-servers, and support groups sprung up around the world. Although full-out warfare with the Mexican military was preempted by the presence of a global audience, the Zapatistas haven't yet won the constitutional re-entrenchment of the right to collective community farming.

In the year 2000, I got involved with a Toronto support group that raised awareness of the conflict and money for the communities in resistance in Chiapas. Through that group, I became involved in a project to restore a small, self-contained hydroelectric infrastructure on an abandoned coffee plantation that had become a refugee community. I wrote a funding proposal for the project; after about a year and a half, we finally had enough money and a three-person team of volunteers to go and work on the project. We packed as many parts as we were allowed by airline weight restrictions into big plastic pails and set off from Toronto. We traveled for two days, by plane, by taxi, by microbus, in the back of a truck, and finally by walking for half a day through the jungle carrying our equipment on our backs. We arrived to a crowd of people who gathered in the village center to look at us, who then, to my amazement, played basketball with us on their hard-packed mud court. How this Canadian-invented sport made it to the center of that jungle village I cannot imagine, especially given that most of the residents I spoke with had no concept that a nation called Canada even exists. I was asked a few times by different children whether I knew how to fire a weapon. In total, we spent three weeks in that community, working every day, all day long. We made their little hydro-turbine operable, rewired the central part of the community, and then removed the drive shaft from the turbine to take it out of the jungle with us and have it straightened in the city.

Figure 5-1. Children of Chiapas. Left: Boys take a break from playing basketball and soccer. Right: Girls wear their best dresses for a picture.

* * * * *

Globalization is said to "open up" national borders to the movement of people. This is not mass migration in the colonial sense in which a migrant population was assembled to invade/occupy specified physical territories and displace native peoples. Rather, contemporary migration on a global scale involves the movement of peoples from an exceptionally broad array of nations and regions—movement from less developed to advanced capitalist nations, the reverse, and economically and regionally lateral migrations. Thus there is great need to understand migration in the present context (Blunt 2007). Although there remain many predominantly monocultural nations throughout the world, a very new "mixed" demographic has emerged in many Western advanced capitalist nations:

> The world, all of a sudden, has become a very crowded place, and communities of minorities and postcolonial immigrants now populate metropolitan schools and suburban towns to an extent and in a manner that deeply unsettles racially hegemonic groups. (McCarthy, 1998, p. xi)

Hardt and Negri describe the mass of people in spatial transition as constituting a "multitude" that has been fundamental to the maintenance of capitalism since its birth (Hardt and Negri, 2000, p. 397). But contemporary mobility continually breaks the traditional boundaries of colonialist capital and is less containable since it is now a fundamental part of productive systems, and thereby the multitude "reappropriates space and constitutes itself as an active subject" (Hardt and Negri, 2000, p. 397). Contemporary production happens in the space of movement and depends on migration and mobility.

Silicon Valley and even specialized districts within contemporary cities that attract scientific experts and technologically savvy populations exemplify what Castells and Hall (1994) have described as "technopoles." Elite specialized labor communities populated with international citizens are one localized economic extreme in the globalization of labor. The second is the resettling of cheap manual labor.

The Latin American Maquilladoras (factory free trade zones) such as those found in Northern Mexico and Guatemala have, in this stage of globalization, become features of numerous industrializing nations. These zones form the backbone of globalized physical production, tending to attract migrant workers forced from their traditional way of life, or who have few or no other choices for gainful employment. This is a more lateral type of internal migration, ranking at the opposite extreme than technopoles on the global job market. In addition to often entailing separation from traditional family units, these manifestations of mass migration also often entail disruptions in traditional cultural roles and values. These are sometimes welcome, as in the newly evolved economic power of many Mexican women who suddenly find themselves in the role of family provider. This is described as a "positive gain of the globalization process" (Mejía, 2002):

> Women are taking an active and protagonist role in responding to the changing economic world and the implementation of economic policies developed through a clear male bias. Their activism includes defending their families, communities, traditions and cultures, as well as developing a gendered analysis of the changes taking place. (Mejía, 2002)

Though there may be cause for some cautious celebration, such economic empowerment comes with the price of long grueling hours of toil and overtly oppressive conditions, entailing great difficulties and sometimes tragic consequences (Ong and Nash, 1994).

Wireless Guerrillas in the Jungle (Scene 2)

Our team leader, Vic, had brought his laptop with him, an old gray IBM. He usually kept it in the shed with his luggage, where we slept on bare wooden planks, but one day he took it outside to do a little bit of writing. The iconongraphy was certainly not lost on me—this was, in practice, their principal technology of resistance. The computer quickly drew a crowd around itself. I got the impression that actually seeing a laptop was, in fact, a very rare and strange thing for those so-called postmodern warriors in that community without electricity. There was the odd solar-powered battery charger outside a few of the dwellings, used for a couple of lightbulbs and to run a small television, but certainly no computers. Refrigeration was something not possible. Being a vegetarian at the time, out of sheer politeness I had been unassumingly eating the white bread ham sandwiches that arrived every day at the work site, sent by an invisible woman. However, when I found out that someone had been walking several hours through the jungle every night to get this prized meat from a refrigerator in a neighboring town, I made them promise they'd never do that again, and that they would serve us the same beans and tortillas that everyone else ate. To me it was obvious that what we were doing was very important. Why else would they go through such trouble to try to keep us comfortable? To document the project and to provide the sponsors some local feedback, towards the end of our stay there I conducted several interviews with the townspeople.

"What is the most important thing this town needs?" I would ask each of my interviewees. The eldest man I spoke to said the town needed variety in its seed stock, because too many crops had been burned in conflicts; they'd lost variety in their diet and that was causing health problems. Another said the community needed potable water, that the nearest source had too much algae in it and so they'd all been hauling water from two miles away every day. Predictably, the community nurse said they needed a doctor to visit regularly and a better stock of medicine, especially for the children's diarrhea. Mateo, who I had become good friends with, said they needed corrugated roofing to build more houses, because their present conditions were too crowded and unhealthy. Another young father said they quite simply needed more food of any type (the townspeople would generally eat only one meal a day). A couple of people commented that in addition to the obvious food and medicine, they needed boots and shoes very badly, especially high boots that stop snakes from biting ankles. I heard horrible tales of violence and death in their communities of origin, of families walking for weeks through the jungle with only clothes on their backs to get to that safe haven, of surviving for two months eating only one tortilla a day, of many children buried in the town cemetery. Every adult I spoke with had lost children from sickness related to malnutrition.

Not one person referred to what we were doing as important.

Not one person referred to what we were doing there at all.

I finally realized the value of our project for them: the village would not be attacked as long as we were there. That was it. The fruition of our toil was a few weeks without fear of violence for the townspeople. I became hyper-aware that with all my relative privilege and wealth, with all my technological knowledge and aspirations to set it to work on behalf of the people, I could do almost nothing to better the circumstances of everyday life in that community beyond simply being there. Even though our small team included a computer programmer and an electrical engineer in addition to my expertise in communications, I'm certain my most lasting and positive contributions to the lives of the people of that community was showing the children how to make whistles with a blade of grass held between the thumbs, and how to make a loon (a bird they are unlikely to ever see or hear) call by blowing across cupped hands. Their squeals of laughter and cacophony of whistles sailing across the jungle are the hope and everyday inspiration of the so-called postmodern revolution. I fantasize of returning there one day and knowing the community is doing well, not by finding it lit up and electrified but rather hearing the children still laughing and playing games with their improvised whistles the way I showed them. Maybe sometimes, just whistling can be revolutionary.

* * * * *

Another form of global migration can be seen in the ongoing U.S. federal debate over "illegal" Mexican workers. While it remains legislatively taboo, the employment of "illegals" for indigent labor in the United States is tacitly approved in practice. Mexican and other willing workers are welcomed to fill low-paying and "dirty" (usually physical labor) jobs that U.S. citizens will not perform. A similar phenomena (though not always illegal) has become part of the Canadian labor landscape. Nannies and housekeepers brought from Asia form a significant demographic of the Canadian labor market. Canadian importation of seasonal farm labor from Mexico and the Caribbean has resulted in worker towns such as Everton, Ontario. Everton is said to reflect a Latin American labor camp in the heart of Ontario's farming region.[1]

There are other examples of physical migratory patterns in the context of globalization (e.g., see Held et al., 1999; Castells, 2000), but the important issue in terms of a cultural view of globalization entails the relationship of the symbolic and performative processes that are disrupted, forgotten, and/or preserved, mutated, and re-articulated in new (and old) locations as a result of globalization. In this age of hyper-electronic mediation, physical migration does not have to be part of the processes of re-articulation. Rather, the technology itself may act as the agent of articulation.

A Standard Arrival in Tokyo

The jet plane taxis down the Narita Airport runway in the gray, misty Thursday afternoon, and cellular phones begin to appear in people's

1 http://www.macleans.ca/topstories/canada/article.jsp?content=20041011_90409_90409

hands as a woman's voice announces something in Japanese. I'm eager to call home and announce my safe arrival, so I take out my Motorola V600 cell phone. I wait expectantly while it powers up, watching the signal indicator. Nothing. I begin to wonder what the problem is. I look around. Other people are already talking on their cell phones. I open the "tools" menu and choose the "search for network" feature. When I enrolled in the international roaming plan, the instructions said I should get a list of providers to choose from when I power up the phone abroad. After a minute of searching, still no networks appear. I overhear a conversation two rows back between a couple of American soldiers:

"I carry two phones with me all the time…'cause you know, the *standard* is different here in Japan, and it's cheaper to have a Japanese subscription and call out with a phone card anyway, instead of using roaming."

His counterpart complies with his technical expertise, "Yeah, I guess I should get another phone too…I don't know how long I'll be here though…You know…If it's worth it…" I and other passengers within earshot find out the former is a technology specialist in the U.S. army who works mainly in Japan, Korea, and Alaska, and the latter is going to join up with a unit at the same base his new buddy is headed to.

"Welcome to outpost Japan, where the *standard* is different…," I sigh to myself soundlessly. Disappointed but retaining some hope, I decide this might just be an "airport" problem and that things might be a little different in the city in terms of network access.

Inside the terminal, I pass through the immigration booth with relative ease, somewhat automatically gaining a 90-day visitor's visa. I claim my luggage from the carousel and wheel it to the shortest customs line. The officer quickly finishes with the woman ahead of me. I drag my two bags with me and hand him my passport.

"Hey…," I say seriously, nodding slightly.

"You are from what country?" he asks me. I assess him while he speaks: late twenties, thin, connective but not friendly nor intimidating either.

"Canada," I say, "but I live in the USA."

"Why do you come to Japan?" he questions.

"Tourism…I'm a tourist…," I say. "Just for a vacation," I tell him. He looks at me calmly then turns to his left and selects a blue three-ring binder from among various other books on the shelf by his computer terminal. He silently opens the binder and deliberately flips to a specific page.

"Do you have any of THIS with you??" he demands with an intimation of aggression, thrusting the book towards me.

"Hmmmph...!!!" I choke a laugh inside my throat but erupt in hysterics inside myself as I look at the picture he has presented to me. It is a collage of pictures showing very large quantities of marijuana set out on display, along with a picture of a quarter-ounce baggie and another picture of a joint and some other paraphernalia. "No," I say, still trying not to laugh. He nods thoughtfully, turning the page.

"Do you have any of THIS with you?" he demands again, this time showing me a page full of pictures of heroin in a similar array.

"No," I repeat with a mechanical air. "Wasn't heroin one of the first globally traded products...?" I question to myself. We continue this surreal debacle of drug interrogation through pictures of cocaine onto those of methamphetamine, ending with illustrations of barbiturates. The phrasing of our interaction remains the same throughout. Finally, he asks me to open one of my suitcases.

I lift my rolling Samsonite duffle onto the counter and unzip it. He pretends to look at something inside it, then says, "*Arigato*...Enjoy your time in Japan..."

I say, "OK," wondering if this comedic screening is the result of Canada's recently attained reputation as the producer of the world's most potent pot.

With luggage in tow, I walk through the door to the arrivals waiting room. I stop for a moment to appraise the situation. It looks to me like most airports I've been to, but with Japanese writing. Immediately noticing a liberal peppering of English throughout, I feel reassured that my lack of a functional Japanese vocabulary shouldn't pose a huge problem for me. I begin absentmindedly walking to my right, in the general direction of what looks like an information booth. I find I am right in my assumption and wait in line for my turn with the clerk. When I approach the counter, one of two smiling young female attendants bows and greets me in her Japanese attendant's voice. I ask unashamedly, "Do you speak English?"

"Yes...a little bit...," she answers in a voice that sounds much more natural to me than the border guard's. I ask where I can buy tickets for the Skyliner Express train to Ueno station, and where I can find a bank machine. She answers each question in turn, pointing in opposite directions to guide me.

"Thank you," I say, pick up a free map of Tokyo, and head in the direction of the bank machine. Passing car rental booths, hotel information counters, and various food stands, I am suddenly aware that I'm passing a row of cell phone rental companies. Just as I'm about to get interested in this development, I spot the Citibank machine out of the corner of my eye. I head over, cram myself and my luggage into the booth, and withdraw several thousand Yen. Extracting myself from the money booth, I wander calmly back to the cell phone rental companies. "Is my phone really not going to work here?" I wonder to myself. "Should I just rent one right now?" T-Mobile told me my phone would work in Japan, so I make up my mind to believe them and stick with my original assessment; I tell myself that in the city it will be different. I begin comparing plans anyway, just in case I decide later that a rental is the way to go. I spot a young Western man doing the same thing.

"Hey, are you going to rent a phone here...? Have you ever rented a phone here before?" I jam the two questions together before even thinking to ask if he speaks English.

"Yeah, I'm going to rent one I guess...," he answers with an east coast U.S. accent.

"Do you know which one is best to rent?" I ask him.

"Well, my dad told me which one to get, you know, but I can't figure out which one it is...he's here now...actually, he works here a lot...I'm from Virginia, and I've been here a few times with my dad, but I never rented a phone here before...," he volunteers.

"Do you think it's worth it?" I ask him, "I mean, do you think it's expensive? I'll be calling overseas, and my incoming calls will be from overseas too..." He affirms that it will be VERY expensive to rent a phone for that purpose.

"I'd suggest getting phone cards and just using pay phones," he tells me. I decide to take his advice and forgo the rental. I thank him and proceed to the Skyliner ticket counter as my first step to finding my way into the city.

I find my way to the platform and after a few minutes board the train to the city. I carefully stack my luggage in front of me, keeping a hand on my backpack with my laptop tucked inside it. I watch others board, noticing that most of the other passengers seem to be sans luggage, and about one in four, both women and men, is holding a cell phone. Filled to standing room only, the train begins forward as I look for the bars on my phone screen to indicate a signal. I watch other people text messaging but don't see anyone actually speaking. I long for my phone to work. I hold it like a friend in my hand, periodically checking the signal indicator bars in hopes that my fear is not true. I turn it off and on again a few times, thinking maybe I just need to reload the memory card. I find the network selector control panel and scan for any available signal. Late in the hour-and-a-half ride, well inside the city and seeing others still using their text-messaging programs, I come to terms with the fact that my phone is definitely not compatible with the Japanese *standard*. I swallow my regret that I didn't rent a phone at the airport. "Should have listened to the soldier…," I reflect. Resigning not to make this long trip back to the airport just to get a phone, I surmise that as far as cell phone ethnography goes, this field trip to Tokyo will be all about observation, and not much participation. I console myself: "At least I still have my laptop! I'm sure I can find WiFi everywhere."

* * * * *

Globalization as an economic project is said to facilitate the trade of goods and capital on a physical expanse grander than ever before, and in a temporal sense unthinkably fast. On a practical scale, "all electronic processes…can be reduced to knowledge generation and information flows" (Castells, 2000, p. 409). But communication makes network society and globalization possible. In addition to the economics of digitization, globalization of digital media facilitates instant communication in everyday life. Global networks create "the space of flows," which materially organizes "time-sharing social practices that work through flows" (Castells, 2000, p. 442). The space of flows is composed of electronic space (waves, circuitry, nodes, and hubs) and the human hegemons that control it. This gives everyday people access to information and cultural icons on an unprecedented scale. However, this unprecedented circumstance comes with its downside. Global economics demands "absolute conformity of all goods without any cultural exception" (Virilio, 2002, p. 63). To not conform means to not participate.

The movement of financial information has inspired incredible advances in computer technology and policy. Digitization of money transfers in the stock exchange and the facilitation of global trade through the Real-Time Gross Settlement Systems (RTGS) used to conduct exchanges between large financial institutions have literally provided the financial infrastructure for international business (Kien, 2004a). The impacts of this digital communications infrastructure is altering the terrain of financial exchanges and have changed the nature of global economics, granting the ability to instantly move large amounts of money to any point in the network at almost any time. New technologies grant more people access to the financial network to participate in global exchanges. Although the global digital network is a material entity in terms of equipment, one of its main impacts is the "dematerialization" of money itself. The global "information" economy has emerged, wherein the transfer of information and its material and knowledge-based infrastructure have become central features of the economic base. The digit of information is at the same time both commodity and counter. In this configuration, domestic markets retain the strongest share of GDP but are profoundly implicated in world trade (Castells, 2000, p. 95). New rules have been invented to keep national and international markets under control in this context. For example, rampant currency speculation raises the possibility of political instability (Crotty & Epstein, 1996). Thus global protocols on transfers now exist as international regulatory mechanisms, a result of international governmental and corporate agreements. States and corporations negotiate with each other to define the terms of global trade policy. Aside from the obvious benefits of global access to goods and resources, the opening of global markets is an important motivation for the corporate world. The diffusion of trade goods on the world market is epitomized by the globalization of culture.

Mr. Yahoo

I'm seated on the subway, mid car, headed to the COEX shopping mall on a Saturday afternoon. How I got here, I hardly know. I remember wandering towards the station after lunch, but now my mind is almost blank with a typical post-lunch energy low—I'm able to think only about where to get a good cup of coffee. I close my eyes. In my afternoon stupor, I'm barely aware of the subway going through its routine of stopping and starting. I open my eyes. Crouched in the middle of the crowded subway, a young man balances his laptop in his left hand and pecks at it with his right, all the while surfing precariously as the subway lurches along its way. It looks insane, recklessly surreal, like he's about to drop it at any moment. "Where the hell am I?" I wonder to myself. I reluctantly force myself out of my drowsing. I want to see what he's doing. More, I feel it's my duty as a researcher to investigate this spectacle. I casually stand up and try to maneuver through the crowd until I'm up behind him, but as I approach, he folds up his computer and tucks it into his black backpack. I watch someone claim the seat I just vacated. The joke's on me. I look him over from this closer range. He could be any normal Korean university student, except that his backpack has a tag dangling from it with a business card tucked into it. The colorful Yahoo! logo broadcasts his IT importance. I can tell it's a real business card. He uses the English name Steven, but since most of the writing is in Korean, I can't figure out what his title is. However, the card is already enough to wink to the world that this otherwise nondescript twenty-something Korean man is a somebody, part of the Yahoo! corporate global alliance, with so many things to do and places to go that he has to work even while standing in the middle of the crowded subway.

CHAPTER SIX

Post-Global Citizenship

One Night in Seoul

"!$%&^*...!!!"

The quite conversation I'm having with my friend is interrupted by a sudden tirade of Korean curses from a small, tanned old man almost right beside us. It's Saturday evening, and we're riding the subway in Seoul, on our way to a movie theater. The old man goes on shouting something in Korean that I can't understand. But Sean does. A timid teenage girl holding an open cell phone in her hand is the recipient of his wrath. I look around the subway car. Some people stare with a bemused grin, while others stare away as if nothing unusual is happening. The girl silently sits, red-faced, looking at the screen of the phone in her hand. My first instinct is to assume they are related, a grandfather scolding his granddaughter for some kind of domestic impasse. When I ask Sean what's going on, I get a different understanding of the situation. He tells me the old man was yelling at the girl because he could see that she was text messaging in English on her cell phone. He was screaming "This is KOREA! WE SPEAK KOREAN HERE!" The old man gets off at the next stop, and the girl is quickly forgotten about by everyone around us. I notice she is still typing into her phone, messaging to god knows who.

"Was it because of us?" I ask Sean ignorantly, thinking our conversation might have provoked him.

"No... no, you hear this all the time once you can understand Korean...sometimes they shout it at you, and they think you can't understand them...usually I just smile at them, but sometimes I answer them in Korean, like, 'What? Did you say something to me, uncle?' You know, something long and really polite so they know I speak Korean..." I mentally record the scene, thinking of how important this is to my ethnography.

* * * * *

The acquisition of social capital doesn't always result in upward mobility (Mitra, 2008). In addition to meeting the criteria for acquiring greater social capital, there needs to be room for movement and subjects must be allowed to move. The relationship of material to metaphysical realms is generally described by Foucault (1979, 1980, 1990) as a relationship in which domination of the body as a physical territory is used as a means by which to instill behaviors (taken as an indication of orthodoxy) that reflect what the dominating power/knowledge considers appropriate. The conceptual paradigm of sacred/near versus profane/apart is reflected in the subjective construct spiritual expert versus heretic. Foucault's aesthetic of the panopticon maps this relationship: experts with power intermediate to form a guard around the spirited conceptual center, enframing all-seeing God in the panoptic center by holding the profane world apart. Those without power are relegated to the profane edges—correlative to a Heideggerian standing reserve—where they are further barred from access to expertise. Expertise is won only through controlled exposure to the center of power, through controlled opportunities to learn and enact the performative cultural script that is held as the legitimate discourse of power. This is the logical use value of the modern human being—to reproduce the cultural order, by being called to learn and perform the social script.

While the globalization of culture often works to the detriment of local concerns due to the exclusion or disruption of local issues and practices, it would be erroneous to simply accept a "magic bullet" or "hypodermic" theory of communications effects:

> It would be a mistake to theorize the global as merely homogenizing, universalizing, and abstract in some pejorative and leveling sense in opposition to a more heterogeneous, particularizing, and concrete local sphere. (Cvetkovich and Kellner, 1997, p. 13)

Neither is globalization entirely benevolent. Nevertheless, in the relationship between communications and local culture, it is the disruption and forced change of local cultural production that are perhaps most problematic. This in itself sows division and confusion: "global takeover of humanity by totalitarian multimedia powers, applying intensively to populations that age-old strategy which consists in sowing division everywhere" (Virilio, 2002, p. 26). "Democratic" political systems become characterized by contests between politics and media, creating a "virtual democracy" conditioned to respond to the "optically correct" rather than the "politically correct" (Virilio, 2002, p. 31). This is the knowledge the Zapatistas were able to exploit so effectively. Globalized communications thus brings "optical challenges" (Appadurai, 2001, p. 3), although with the power to regulate and enforce media as business, content, and consumption, states retain their basis as the nation's fundamental power.

Globalization of Citizenship (Scene 1)

It's summer 2004; June 28 to be exact. On the last night of the four-day International Crossroads in Cultural Studies conference, in a small town surrounded by cornfields in the middle of the state of Illinois, a group of Canadians converge in the Peterson Room of the Institute of Communications Research, University of Illinois. In the Canadian BYOB tradition, they arrive with beer, wine, and drinking snacks such as chips and pretzels. I have set up an LCD projector and have hooked it up to my laptop streaming a live CBC broadcast of the Canadian federal election results. Over the next several hours, this group of Canadian cultural studies scholars transform the ICR into a Canadian space, telling Canadian jokes, talking about Canadian politics, cheering and booing the election results on the screen as if it were a hockey game. Perhaps the scene is not such a stretch to envision, as I had once been told McLuhan had assembled the chapters of *Understanding Media* in that very building. We take particular delight when the Marxist-Leninist candidate from central Toronto briefly takes the lead in his riding. One spectator, Bob, comments, "Well, we know he isn't gonna win, but it's great just to hear Peter have to say Marxist-Leninist on the CBC, eh?" Everyone laughs, a little bit giddy. The occasional comment is hurled at the screen, and as the night wears on, we begin to compare notes on our roots. I find one of my companions, Ken, is from Manitoba, the province I was born in. I'm convinced from his appearance that we're related somehow; I could easily mistake him on the street for one of my uncles. We do a thorough tracing of our lineage, seeking a family connection. We can't find one, but since we both know we have hundreds of relatives in the Canadian west, we agree to suspect that somewhere in the history of our land we are related. Some people come and go. A group of Australians walks in at one point and gets a quick lesson in Canadian political history and, with drinks now in hand, stays for the duration. One of my Canadian classmates shows up later in the night. The Liberal party wins a small minority. Satisfied that there is nothing more to learn from the broadcast, the party slowly breaks up. Everyone walks out with their garbage in hand saying, "Thanks—see you later." Completely exhausted after four days of conferencing and a late-night election party, I pick up my laptop, take a last look at the now darkened LCD projector sitting on the table, and close the door behind me.

* * * * *

Virilio prophesized "Tomorrow's war will be *globalitarian*, in which... the qualitative will be of greater importance than geophysical scale or population size" (Virilio, 2000, pp.144–145). As if to prove him correct, in the wake of the 9/11 attack British PM Tony Blair declared that the contemporary world is in a war over "universal values" (Tony Blair in Virilio, 2002, p. 43). By "universal values" he means, of course, the global struggle for the imposition of Western liberal ideals, at the expense of the displacement and erasure of local and "other" cultural values and systems. This is part of setting things in motion, the mobility of globalization that includes people, goods, communication, and economic and cultural capital. However, to set something into motion, there must be something that appears relatively stable to contain and/or control it: "apparent stabilities" are "usually our devices for handling objects characterized by motion" (Appadurai, 2001, p. 5). Appadurai further contends:

> The greatest of these apparently stable objects is the nation-state, which is today frequently characterized by floating populations, transnational politics within national borders, and mobile configurations of technology and expertise. (Appadurai, 2001, p. 5)

The physical borders of national territory are transgressed in migration, but national borders are even more porous when it comes to communications and cultural icons and practices. Uncommodified ideas are ultimately more portable than goods, and even bodies: "the imagination as a social force itself works across national lines to produce locality as a spatial fact and as a sensibility" (Appadurai, 2001, p. 7). Appadurai's reflections describe two aspects of national territory theorized by Maurice Charland, highlighting that the physical/cultural spatial split that Charland theorized is not just a Canadian phenomenon: "all over the world, the processes of migration, electronic mediation and the work of the imagination of the great masses of the people have effected the separation of culture from place" (McCarthy et al., 2003, p. 13).

Describing the evolution of the physical territory of the nation, Gilroy tells of national "camps" based on military models emerging during modernity that find affinities in articulations of "race" identification within respective nations (Gilroy, 2001, p. 82). These camp-style nations constituted difference on and imagined blood-based ethnic purity (i.e., "royal" blood, etc.), which maintains some cultural currency in the present. Those within the nation without adequate "purity" are perceived as invaders, not belonging, while a perpetual internal versus external conflict legitimates the national camps (Gilroy, 2001, p. 84). The battle is made meaningful in terms of identity and is thus mainly ideological—a fight in cultural space for placement in physical space. While "purity" was always already a campfire myth, the realities of contemporary global migration make the fact that the nation is in practice not a very good container, and that neither are its citizens easily contained, ever more difficult to ignore.

The contemporary nation must come to terms with the cultural reality that the nation is actually *inter*national—it is always already globalized. The "other" is always within (perhaps *is*) the populace, and the borders that would serve as containers are constantly transgressed (Bhabha, 1990, p. 4). "We" may not be who we're told we are: "the stranger, the foreigner, is not only among us, but also inside us" (Morley and Robins, 1995, p. 25). The traditional nation, acting as a container of people, is an articulation of power (Bhabha, 1990, p. 292), whereas the self-consciously re-imagined nation understands that the national "space" is inside its citizens (Bhabha, 1990, p. 297). Nationalism can be a flag one may or may not choose to wear, depending on the opportunities one's context provides. The culturally globalized nation acts not as a container of national identity, but rather as a cultivator thereof. Partha Chatterjee (1993) discusses this process as a result of the imposition of an imagined community via colonialism, and as the natural evolution of a re-imagined "us" according to the reality of a citizenry's circumstances. Anticolonial nationalism grows out of the spirit of the actual people that borders pretend to contain.

If the nation is thus culturally reterritorialized by recognizing individual citizens as the producers of meaning, it seems reasonable that national conflict must then be embodied in individuals (Hansen, 1993, pp.183, 206). This recognition, heightened by the advent of electronic global unification, necessitates the redefining of local and global structures, as publics ebb and flow according to the tides of internal and external international issues. To paraphrase Rosemary Coombe and Jonathan Cohen, it is not possible to define or limit a "proper" realm or range of political activity: "National borders are mobile and diffuse as immigration officials gain access to workplaces in the United States and US Capital interests penetrate even further into rural Mexico" (Coombe, 1998, p. 182). The point here is that the national borders of cultural territories reside in and move with the bodies of the populace, and being such, nations in the context of globalization can no longer be considered simple containers of populations. Nationalism in cultural territory is a constructive process of identification rather than containment.

Globalization of Citizenship (Scene 2)

It's now three years later. The informal coalition that kept the Liberal government in power since 2004 has fallen apart, triggering the first winter election Canada has staged for several decades. It promises to be a tight race, with the two leading parties running neck and neck in the frequently updated polls. With fond memories of the last election, I decide to try for a reenactment. I post a question on the UIUC Club Canada list server: "Anyone interested in getting together to watch the election results online? CBC will stream their broadcast on the Internet." I get a few takers, but not enough interest to warrant booking a special venue. Michael, our president (we have joked before that he should really be a prime minister), comes to the rescue, offering to host a little party at his house.

I arrive at Michael's house with little expectation. I've already convinced myself that no matter which party wins, not much will really change in Canadian politics; either way, it's almost certain to be another minority government with the leftist NDP holding the balance of power. I think of this as mainly a social event, just a chance to be Canadian for a while. Bearing the ultra-Canuck gift of a six-pack of Molson Canadian beer, I reach Michael's doorstep and I am greeted by Jasmine, his German sheperd. I quickly realize Jasmine is not bilingual, but rather a proud but enthusiastically friendly francophone. "She seems to have a lot of joie de vivre," I say to Michael. "Yes, she definitely has that...," he says, then tells her in French to back off of me. I'm introduced to Grace, another Canuck from Ottawa who is already there. We try to set up the download stream with CBC, but it keeps failing. About half an hour later, we are joined by Jean, another Montrealler. After several unsuccessful attempts to get the stream operating, Michael abandons his computer for a few minutes. I'm rotated into the tech chair, and I find I must upgrade his software to make it happen. Finally, new software installed and Michael safely back at the controls, we get the stream to work.

The broadcast is reporting an early lead by the Conservative Party over the Liberals. Jean and Michael seem pretty concerned about the results. "Well, I guess it won't surprise me or even matter that much to me one way or the other...," I suggest, thinking I'm being consoling.

"Well, I don't know about that...," Jean says. "I mean, you think about it...Stephen Harper doesn't really consider gays real people...," he expresses his fear of what a right-wing agenda could bring, referring to the Conservative leader's promise to reopen the gay marriage issue with a free vote in Parliament.

"Yeah, and not to mention that he has already expressed publicly his disdain for Quebec...He hates the French...," Michael adds.

"Oh, yeah, of course, that sucks," I say apologetically. "But I don't think they can really change anything at this point...," I try to reassure. "Something like that could bring down a minority government...," I analyze.

"Yes, maybe...," says Michael, "but I'd just as soon not have to go through that."

"Of course," I say.

In the course of the evening, there, in fact, are no further surprises. The results remain largely the same as when we first tuned in, and by the time the election is called and the new prime minister makes his victory speech, Grace has already left. Jean is napping on the sofa, and Michael and I are staring vacantly at the screen. I rouse myself, "OK, time to go..." My activity wakes up Jean, and we leave at the same time.

A flurry of predictions and analyses fly about in the Canadian media for the next couple of weeks. For the most part, it is predictable stuff, and it seems to be business as usual in Canadian politics. We all seem to have adapted rather normally to the election results. Not so for Michael's computer.

Two weeks after the election, I read an email from Michael pleading for help. It seems the latest version of Windows Media Player (the software I downloaded to play the election on his computer) won't allow any language but English in its menus. It keeps changing his French and Spanish music titles to English gobbledygook, dominating his cultural interface with anglophone hegemony. "Seems like the Conservatives got inside his computer that night, and just won't leave...," I joke to myself but, then, reflectively, feel some remorse for being the anglophone who unwittingly commandeered his system. "Such are the subtle workings of hegemony...," I philosophize, turning my attention to the next email in my inbox, which is a call for papers called Canada/Migration/Discourse.

Figure 6-1. CBC Online. Streaming news broadcasts keep the mobile citizen connected to home.

* * * * *

In the contemporary context, network has come to metaphorically characterize all human space: "There is no longer a place that can be recognized as outside" (Hardt and Negri, 2000, p. 211). Subjectivity is constructed in networks: "no subjectivity is outside, and all places have been subsumed in a general 'non-place' " (Hardt and Negri, 2000, p. 253). The nation as cultural territory is a culturally spatialized network. However, Cvetkovich and Kellner (1997) point out that macro-perspectives such as this, accurate in some ways as they might be, tend to present totalizing and reductionist accounts of globalization as a singular phenomenon. Analysis of localities and particularities of everyday life present a more accurate portrayal of globalization as it is actually experienced. This allows us to "see how local forces and situations mediate the global, inflecting global forces to diverse ends and conditions and producing unique configurations for thought and action in the contemporary world" (Cvetkovich and Kellner, 1997, p. 2). This linking of global to local entails the linkage of nationalism to identity.

Postmobile Identity (Scene 1)

"What the @#$%!!... ." I stare at the final box of the online application form, which demands a State of Illinois driver's license number, or alternatively an Illinois State ID. "Can't get even a new phone without becoming American...," I criticize to my furniture. Who would think buying a new cell phone and carrier would entail such a crisis in identity? After exhaustive research online and by word of mouth, I've finally decided on a T-Mobile package sold on Amazon. As an authentic Western consumer, I consider the loss-leader of the phone to be my right and calculate that I will actually earn $120 after rebates with this choice of package. I could get a larger rebate if I chose a different phone, but my aesthetic sense of status won't allow me to get anything less pleasing than my present choice. Trying to finally place my order this warm February evening, I'm suddenly stuck at the credit information section and can't move on. Already predicting the outcome, I try anyway, punching in the numbers on my Ontario driver's license. The error message pops up: State ID invalid. Although I was advised by numerous people that a State ID is useful, I've resisted this hegemonic intrusion into my identity until now. In fact, it hasn't been a problem until this moment. I decide to try contacting T-Mobile to see if I can get around this.

I dial the number, and my call is answered after two rings. The female voice of their automated response system begins asking me questions: "How may I help you? You may choose any option from the following menu, or you may say 'representative' at any time to speak to a representative..." The machine begins listing my options, and I start trying to navigate their system. After a couple of mistaken interpretations on the part of the robot attendant, I figure out the proper machination of my voice, such that the switches take me to the right place in their database. After a considerable trail of automated answers, I'm finally told, "OK...let me transfer you..." I hear the machine click. A few tones like that of an old dial-up modem sound in my ear. Suddenly, a different automated female voice comes on the line to apologize and tell me, "I'm sorry, I cannot complete your call right now. Please try back later." AAAAArgh! Maybe T-Mobile isn't the best way to go after all? I decide to try one more time.

This time, I call a different number. I embark on a similar process, but I realize I can cut off the voice with my answer any time I want. Of course, this is a rather rude way to proceed for a Canadian, even if it is just a robot I'm dealing with. Conventions aside, I quickly find myself hearing the now familiar dial tones, and my call is successfully transferred. After a couple of rings, a different female voice tells me I have to wait for a representative. I listen to some music for a minute, and then a man who tells me his name is Paul answers. I start to tell him my problem. He tells me that I have to talk to a representative in a different department. He tells me he'll transfer me himself. I tell him OK, as long as I can talk to someone. He then gives me the same number I just dialed, "just in case you get cut off..."

"This is actually the number I just called, and I got you!" I tell him.

"Oh...Uh...I'll tell you what...I'm going to 'conference' in on the call, so don't worry about it, I'll take you there myself," he assures me and tells me he's choosing menu item "0" to get a representative. A moment later, a woman answers. I thank Paul, who leaves the line. The woman, Julie, listens as I explain my State ID dilemma, and I ask if there is any way around this on the web site. She tells me she was born in Ontario herself, then puts me on hold and begins investigating things from her end. She makes several calls, checking back periodically with me. Finally, Julie the Canadian informs me that she's found the right person. She says goodbye and unceremoniously transfers me to a male salesperson who confirms there is no way to get around the form because it's a special deal on the Amazon web site only. He tells me I can go to their store and sign up through T-Mobile directly if I want to. I explain that I wanted the deal on Amazon. I thank him and hang up. I resolve that I need to get a state ID. I've been swayed by money. "So American...," I say disapprovingly of myself, as if saving money were a wholly American ideal.

* * * * *

Identification is a way of understanding and making meaningful everyday cultural territory. Its popularity as an interpretative device is helped by its plurality of meanings. Sharing identity gives people a feeling of bondedness on a very deep level. However, Gilroy points out, "Nobody ever speaks of a human identity" (Gilroy, 2001, p. 98). Human identity, rather than a more specific sub-identity, "orients thinking away from...basic, anti-anthropological sameness" (p. 98). Identity is culturally specific, not necessarily associated with race or gender, but most importantly it allows us to speak of "we" and "they" as political collectives. It is important to understand the processes of identification, since "the idea of fundamentally shared identity becomes a platform for the reverie of absolute and eternal division" (p. 101). Uniforms and other "spectacles of sameness" may be employed (p. 102), and identity becomes a "thing to be possessed and displayed" (p. 103). Obedience to an essential identity signifies belonging. However, to be mixed is to present a confounding betrayal (p.106).

As mentioned, globalization proffers pluralism. Globalization and electronic mediation has transformed the nation-state to now consist of more and more hybrid-identified citizens, identifying themselves in terms of multiple nationalities. However, in the dominating structure, citizenship is positioned oppositional to transnationalism. There is a paradox in this, in that one must first become national in order to become *trans*national, but it is loyalty to *one* flag (one tribe as it were) that cannot be transgressed without it being considered betrayal. Gilroy (2001) points out that giving up notions of essentialized identities faces fierce resistance in all quarters, since they are often seen as a source of pride among people. Identity is considered a means of acquiring certainty about oneself, and subjectivity and consumerism are a means of constructing one's own individual identity. Identity is also constructed by "difference" (external relationship—"othering"), which is the inverse of identity by "sameness" (internal relationships) (p. 109). And, of course, notions of belonging—"place" or "placelessness" (diaspora)—may influence national identity. Even identities as historical "victims" have political currency in some contexts (p.112).

Emphasizing individual and national identification is a tactic of resistance to globalizations, but globalization also brings about hybridized and new localized identities (Cvetkovich and Kellner, 1997, pp.9–10): "although global forces can be oppressive and erode cultural traditions and identities they can also provide new material to rework one's identity and can empower people to revolt against traditional forms and styles to create new, more emancipatory ones" (p.10). As a process, identification is another way of mutating, combining, and rearticulating old cultural practices, and articulating brand new ones: "even individual identity is more and more a question of articulating often conflicting cultural elements into new types of hybridized identity that combine national cultures with global ideas and images" (p.10). However, a problem arises in this type of celebration of identity, as it tends to treat "traditional" identities as if they were fixed/static, or always desirable in the first place.

Postmobile Identity (Scene 2)

The next day, I borrow my friend's car with squeaky brakes and head to the DMV in the bright morning sunlight. In the general chaos that is deemed the waiting room, I get in the wrong line. A big white man redirects me to the info line. After waiting about ten minutes, I finally get up to the counter. A tall blond woman asks whether I have an out-of-state driver's license. "Yes…," I say.

"You will need to give it up for a State of Illinois ID to be issued," she says.

I'm dumbfounded. This doesn't seem like a fair trade at all! "Really??... I'm Canadian…," I stammer.

"Oh, that's a FOREIGN license then…," she distinguishes between out of state and out of country. "You can keep your Ontario license then," she consoles me. I present my passport, social security number, and my Ontario license. She glances at the pile of information, then tells me I need something with my Illinois address on it. I realized I don't have any such thing on my person. Telling her I'll be right back, I go out to my backpack in the car and retrieve my checkbook. I cut directly to the front of the line so she can verify that it's acceptable, and she redirects me to another line snaking through the lobby.

This next line moves at a snail's pace. There are only two people—a middle-aged white woman and an elderly black woman—working at the counter, leaving another three workstations vacant. After over 20 dazed minutes in line, I realize I'm going to be late for work. The irony of the situation strikes me as I take out my cell phone and call the office, explaining "I'm at the DMV," as if the problem will be obvious. Apparently it is.

"Ha ha ha…Yes…good luck there…," James replies, laughing knowingly. Another 10 or 15 minutes later, I'm finally at the front of the line. By now the white woman has disappeared into the back somewhere, leaving her aged black comrade to manage the front line by herself. I walk happily to her counter space, place my documents on the counter, and tell her I want an Illinois state ID.

"How are YOU doing?" she asks.

"Good…I'm just glad to get to the front of the line!" I say.

She rather automatically starts handling my papers, bantering: "Well, good…I tell ya, I could write a book about all the things I've seen in here…" Unprovoked, she tells me she's been working there for 28 years. In the course of our interaction, she asks me my birth date four times, the final time explaining, "This darn computer's not working right..." We finally finish, and with the form she just gave to me in hand, I move down the counter to the cashier.

I stand silently for several minutes in front of a huge middle-aged black man, listening to him breathe almost asthmatically as he enters something into the computer. Finally, without looking at me, he takes my form with a 20-dollar bill attached, punches some buttons on the keyboard, and orders me to take a seat. I weave through the waiting room, around the perimeter of a roped-off area with a sign that says "Written Driving Test Area" where there are a few people sitting with some paperwork in front of them. I sit along the back wall. After several minutes, a nervous-looking young black man sits next to me with a seat in between. After several more minutes, I say absently to him, "Seems like nothing fast ever happens in here, eh?" He looks at me for half a second, looks away, and says dismissively without any expression, "yeah...sure..."

I take out my cell phone and play Bejeweled for a while. Finally, the man who doubles as cashier calls my name, orders me to take a seat so he can take my picture, snaps a photo in front of the blue screen, and (still without looking at me) tells me to sit down again. "The only image he has seen of me is what he framed in the camera...," I ponder uselessly, mentally passing the time. "He is a machination...," I condemn him to the Heideggerian standing reserve. I watch as he busies himself with painstaking slowness with some kind of paperwork beside the camera apparatus. He calls another person, a middle-aged white woman, and takes her picture the same way. Meanwhile, I watch what I assume is my finished ID card slowly slide out of the machine and into a wire basket. I stare at it impatiently, glancing every few moments at the man behind the counter who refuses to acknowledge that anything has happened. Indeed, he steadfastly refuses to look up from whatever is in front of him at all. After another five minutes he finally reaches over, takes out the card, and then barks out a mispronunciation of my name. I spring into action, grabbing my coat; receiving the card that he hands off to me, I hastily make my way towards the side exit. I stuff the card into my pocket without looking at it.

An hour later, I'm safely at my desk back in the office. I take out the card and give it an examination. Someone held their keystroke too long and put an extra "3" in my street address. Even so, I'm now officially identified with the state of Illinois, even if my residence is said to be 3,000 blocks away from where I actually live. But it doesn't matter, it's not where you live that matters, it's where the government says you belong that makes the difference. That night, I try ordering the phone again. This time I am able to complete the online application, but, without thinking, I choose to pay with my Canadian credit card and type in my Canadian billing address. The next day, I receive an email telling me I need to use a U.S. address only. Finally, two days after placing the order, confident that I am firmly ensconced in the United States, Amazon and T-Mobile accept me as their customer.

* * * * *

The tenacity of essential identification may be useful in some ways but is certainly not welcomed by all. Globalization has brought with it what McCarthy calls "the new essentialisms, infecting discourses of culture and knowledge" (McCarthy, 1998, p. xii). Alarmists suggest there is potential for various crises of identity. Baudrillard points out that this is a distinct crisis directly related to globalization: "This is not... a clash of civilizations or religions," but rather "triumphant globalization battling against itself" (Baudrillard, 2002, p. 11). With increasing hybridity but tenacious propensity to condemn the "mixed" identity, "each of us is a criminal going unnoticed" (Baudrillard, 2002, p. 20).

Such discussion is problematic for moving forward with an acceptance of the present circumstances, because it reifies romantic notions that identity was at some mythical time "authentic" and "pure," and that when cultural artifacts and practices meet there is always a contest for domination that creates liminality. It also tends to impose a Western view of "multicultural" and/or hybrid globality onto the rest of the world's nations, many of which are dramatically mono-cultural in contrast: "The tendency to treat the topics of culture, identity, and community simplistically is one of the weaknesses that mars current writing on multiculturalism" (McCarthy, 1998, p. 148). Distinct cultures are durable in time and space, adaptive, and thus able to disseminate and assimilate new practices and artifacts, to gain as well as lose some without losing their own identity. For example, the "fortune cookie" is a contemporary Western, not Chinese, tradition. And hidden cultural boundary maintenances occur in the most banal and silent exchanges (e.g., the Chinese language menu silently presented to Chinese speakers in Western Chinese restaurants is often an entirely different selection of dishes, completely unknown to non-Chinese speakers because it is never offered to them).

Consumption of cultural products in diasporic contexts does not in itself indicate a shared culture—culture requires a great deal more synchronicity between people than the "quantitative" experience of consumption and production (Polan, 1993, p. 35). It would be misleading to say that there is a global *culture* when in practice there are merely shared experiences with particular broad phenomena. It might be more accurate to say that globalization takes on the expression(s) of the participating culture(s). Culture and nationalism may flow through each other, but ultimately globalization cannot be described as one single broad cultural phenomenon the way that nationalist expressions might be. But globalization does increase the potential to bring cultures into contact, and thus to augment pluralism. Diaspora necessitates a reproduction of "sameness" that is not the same, but rather hybridized (Gilroy, 2001, pp.129–130). For example, a notion of African identity is preserved through generations throughout the Americas and Carribbean. As Gilroy points out, Bob Marley is African identified but was at the same time a "global" citizen.

Globalization's effect on culture may be understood to be part of its nature as a phenomenon of "complex connectivity" (Tomlinson, 1999, p. 1). Changes to the nature of "local" life occur as a result of intercultural exchanges (Tomlinson, 1999, p. 9). Specifically, having a "global" perspective is normalized as part of one's cultural paradigm (Tomlinson, 1999, p. 30). The result is a disruption of the local relations, as an understanding of global relations takes shape. In a Certeauian analysis, the individual moves through a given territory according to their own logic. Appadurai seems to concur:

> Political subjects are not mechanical products of their objective circumstances... the link between events significantly separated in space and proximate in time is often hard to explain. (Appadurai, 2001, p. 5)

Identification is a process, not an essence (Gilroy, 2001, p. 132). Identification is not rational, and "authenticity" is something manufactured, including "hybrid" nationalist authenticity, be it in physical or cultural territory.

One Night in Quebec City

It's a pleasant summer evening in Quebec City. I'm buying some fruit and a few supplies at a convenience store. "Je ne parle pas français très bien...," I say to the young man behind the counter. "Just a few words...," I continue.

"C'est un sac, oui?" I confirm, pointing to the bag in front of me.

"Oui…An English dat wants to learn French is very rare…," the blond, short-haired Quebecois store clerk tells me.

"Oh?" I say in surprise. I realize that I've temporarily broken the traditional mold of "Anglophone." "Really? I thought everyone here would learn some French…"

"No no…very rare…," he tells me again with a genuine smile—the kind of smile I think of as authentically "French Canadian."

"Hmmm…well…Merci, bon nuit," I stammer to him, suddenly feeling myself crowded out by a few people I first assume are Pakistani tourists but then reconsider as probably Canadian when I hear them speak what sounds to me like perfect Canadian French. I smile at myself inwardly in disapproval upon realizing my discriminating assumption.

"Ciao," he says with a laugh at his little language joke, as I walk through the door into the beautiful Quebec City summer night. I laugh too; "Joie de vivre…," I quietly whisper, unconsciously trying to sound out the words the way they're meant to be spoken.

Although I look, I have the same failure locating an Internet hotspot in Quebec City as in Montreal. I resort to watching CBC and local news en Français in my hotel room. Once again secure in a semblance of my familiar network of mediation and performativity, somehow I feel that here a connection *is* available, if I permit it in the terms it is offered. Of course, there is no guidebook telling you how to discover and navigate the processes of connection, so…trial and error…one script after another, like a hacker trying a list of passwords to break into a computer. I turn my attention to the TV, where the news in French is reporting on a standoff on a native reserve somewhere in the province.

CHAPTER SEVEN

A Theory of Home for the Mobile, Globalized Citizen

When people ask, I say I'm from Toronto. Toronto is my home. Sort of. The fact is, I have been a nomad most of my life—or, more poignantly, a "vagabond" (Bauman, 2000a). In due course, I purchased my way into the more comfortable definition of technologically "automobile" (Hay and Packer, 2004). The acquisition of a wireless laptop and global-roaming cell phone while studying, working, and researching "abroad" thrust me into this definition of subjectivity. My identity border-crossing into the realm of the "tourist" (Bauman, 2000) and beyond has been accomplished not only to the extent of my ability to labor wherever I am. It is also determined by the degree to which my gadgetry allows me to carry the comforts and securities of home with me and, in effect, perform the rituals that maintain my identity wherever I go. As the everyday wireless communications infrastructure of my world becomes more globally ubiquitous, the more transportable the objects of my daily rituals become; my potential actor-network continuously expands. As my objects' portability and wireless connectivity increase, so do the places I can perform my daily rituals. The performances of "I" thus become ever more mobile.

Having already lived in at least 32 different edificial structures I have considered "homes" in the four decades comprising my life, it might be hard to imagine how one might come to envision becoming even more transitory. The issue is not transience, but rather stability. In actor-network terms, it is the ability to maintain network stability while in transit that makes "me" as a coherent entity more mobile, and perhaps less mutable—the stabilizing of "I" into an "immutable mobile" (Law, 1986) to whatever degree I choose. Although I have "taken up residence" (Carey, 1989) in many different locations, it is only in the stabilization of certain daily rituals with the help of mobile devices that I have begun to feel that my sense of identity is better able to withstand such radical uprooting. It makes being a Torontonian easier wherever I am.

Mobile devices enable the transport of everyday performances that maintain a sense of self, which stabilize one's actor-networks, translating

Figure 7-1. Toronto Icons. Top: The CN Tower—the former tallest building in the world—performs double duty as both the dominant cultural icon of Toronto and a functioning communications tower. Bottom: A historical Canadian icon, the Hudson's Bay Company building, sits at the intersection of Queen Street and Yonge Street.

into the maintenance of ontological security even when transience on a global scale becomes an everyday norm. Home as an identified place of belonging is a constantly mutating network of relations performed by humans and nonhumans alike. Technology is a self-mutating aesthetic (Rutsky, 1999, p. 20), a constantly maintained "assemblage of practices... dependent on and instrumentalized through a broad array of practices and technologies" (Hay, 2001, p. 211). So then, how is it that one "becomes" a Torontonian in the first place?

Driving Home

Like the reasonable teenager I was, I decided to move to Toronto to become a rock star. It was 1986. I had quit my job pumping gas, packed everything I owned into my tiny blue Plymouth Horizon hatchback, and drove for three leisurely summer days from where my family was living in Saskatchewan (our family's 8th house since I was born). As if guided by the invisible hand of the city itself, I unknowingly navigated through the highway system to the lakeshore, where I inadvertently turned north. I was completely lost. Sitting clueless at an intersection, marveling at the cleanliness of the late evening streets, I looked up and read the signs "Queen Street" and "Yonge Street." Suddenly, there I was at the most famous and magical intersection in Canada, 18 years old, sitting in my little blue car full of stuff without any idea of what I should do or where I should go next. But, strangely, I felt completely at home, as if I had always belonged there. It was like Toronto was saying to me: "Here...you belong here...right in my heart..."

* * * * *

Of course, even if one might momentarily take up residence there, only insects and rodents can actually "live" at Queen and Yonge; like an airport, it is a purely transitional space. However, the need to construct a place called "home" with theory is not just a telling condition of my own psyche, but also a very Canadian performance in the need to question and construct a unified identity from the milieu of global influences that fragment our national imaginary. Yes, these fragmentations may tend to be overwhelmingly blamed on the United States at present, but lest we forget the British and the French, the vast severity and isolation of the land itself, and the imprint of what "we" call "diversity" from people "we" like to call "native" and "new" Canadians.

While it's easy to get cynical about lip service to diversity, sincere commitment to diversity and the open flow of cultural goods are two key expressions of cosmopolitanism (Woodward, Skrbis, and Bean, 2008). Canada, much like the intersection of Queen and Yonge, is a mobile nation, built and maintained through the national and international mobility of people. This is exactly the kind of labor and cultural mobility that captures the attention of globalization theorists such as McCarthy and Kellner. Globalization may be considered "the intensified and accelerated movement of people, images, ideas, technologies, and economic and cultural capital across national boundaries" (McCarthy et al., 2003, p. 7). But this is no simple, one-sided process. Globalization does not simply absorb the particularity of the local into an abstracted, homogenized universal plane (Cvetkovich and Kellner, 1997). Rather, in effect, the "global" is *always* experienced as a contingent locality, just as many "localities" constitute the substance that is thus "globalized."

Although almost any Canadian city is as good as another to represent the hybrid Canadian demographic, Toronto is the largest and most desirable destination for many, with countless well-defined, government-funded community centers that ensure the adaptation, survival, and ultimate interweaving of formerly "exotic" cultural practices into the "official" Canadian multicultural mosaic. My local neighborhood infrastructure includes Greek, Latin American, Brazilian, Portuguese, Italian, Ethiopian, Eritrean, Korean, North Indian, Pakistani, Chinese, Caribbean, and British establishments, remnants of a dislocated iconography performed with new meaning in the present. These are not simplistic identifications, and their beneficial coexistence is constantly maintained with careful attention: "local forces and situations mediate the global, inflecting global forces to diverse ends and conditions and producing unique configurations for thought and action in the contemporary world" (Cvetkovich and Kellner, 1997, p. 2). Even a short participation in the everyday life of my neighborhood reveals all of these 14 national/ethnic descriptors as solidly Canadian, concerned with life where it is happening in the moment, in spite of the abundance of vague references to far away places. As Boschken writes, "the global city as an interrelated complex system and none exists without considerable questions of relevancy and empirical authenticity" (2008, p. 4). Such references themselves become part of the practice of being Canadian, part of the ritual of serving North African food to Canadian people born in Asia, or the routine of the Latin American grocer selling bagels to a third-generation Italian Canadian. By extension, this becomes part of a Torontonian's culturally performed logic even when away from the city.

A Glimpse of Reality

"Hmmmm...," I mumble to my brother, looking at him desperately. It's morning... time for coffee and the *Toronto Star.* Coffee...one of the most important international trade goods...Cause of untold human suffering, and at the same time, fundamental to my pleasurable daily ritual. The daily news...not just any, but always the *Toronto Star,* no matter where I am in the world, reporting what's happening in MY city...the city that I belong to, and thus belongs to me. My brother Garth knows me well and as a morning person has already made a pot of coffee and had two cups himself. I don't know how we can be so different that way, but he always makes me feel like we belong together anyway. Though I never tell him, he is my hero, having survived a critical brain injury that paralyzed half his body and ended his career as a Jazz musician. He took it like a saint: I never once heard him complain about his situation. On this morning, he nonchalantly presents me with a mug of my addiction. I mutter a distressed "thank you" and walk to the living room with my laptop under my arm. This morning there are three or four Internet signals I can pirate from this second-floor sanctuary overlooking Bloor and Ossington. So many memories in this living room...So many Toronto musicians have called this large old apartment home, so many jam sessions...I take a sip and turn my attention to the *Toronto Star* that has by now loaded in my browser.

* * * * *

Nationalist discourse and its dimensions of citizenship, labor, governance, culture, and ethnicity do tend to dominate the enframing of how we make sense of much of everyday life. However, the nationalist dimension is only one actant of many in the various actor-network arrays translated as home. Nationalist hegemony in social networks tends to provoke an erasure of the more subtle nuances of how we make home meaningful, be it the family or even the national territory itself, apart from the rhetoric of the nation. There is, of course, a notion of home as a place where one hangs their hat.

Home is both a construct in discourse and a material place (e.g., a house), a site of a convergence in time and space that brings together the physical and the conceptual (Morley, 2000). The relationship between the material and the conceptual home is inherently dialectical, in which the conception of home may be equally informed by, and affect change in, the physical characteristics of the home. Actor-network literature describes this as "relational materiality," with form understood to be an effect of relations, made durable and fixed through performance (Law, 1999). Home is an actant, a locus in various networks of translation. In the "discourse" network (typically enframed by a notion of ethnic nationalism), one locates the family, the relationships, and the other socially oriented aspects that go to make up the myth of the home. In the "house" network (typically enframed by geopolitical mappings stabilized through routinized practices), one locates the physical structure, appliances, and the material activities that make manifest the tangibility of the home; in modernist terms, the house is "a machine for living in" (Harvey, 1989, p. 32). The home is a set of technologies and relations that act to conceptually give subjects purchase in time and space, as parent, child, sibling, or whatever other identity performed within the contingency of circumstances. This is no less so with portable technologies, which allow automobile subjects to take with them some of the appliances that comprise the home as a machine, to do the same type of ideological work as the geographically fixed home. Through mobilizing some of the artifacts of everyday performance, one may continue to obtain the intended effect of the ideology informing the construction of the physical space of home—that is, to stabilize one's construct of home even in transit. In effect, in the age of mobility, home is a performance you take with you. For example, I take with me a driver's license that says my home is Toronto. The act of bearing it stabilizes my categorical belonging, and the act of forfeiting such an artifact has the potential to translate into an ontological crisis.

A Typical Morning

"MY SHADOW'S THE ONLY ONE WHO WALKS BESIDE ME...," my clock radio screams beside my head. I automatically reach over and smack the snooze button.

... ...

A while later,

"I WANNA KNOW WHAT IT'S LIIIIIKE...ON THE INSIDE OF LOVE..."

My arm flies out to land on the off button this time. The only thing that makes me move faster in the morning than a song I can't stand is a second song I can't stand. 9:07 am. I fight through the dense morning fog of my mind to contemplate the time. "Better get up now...," I sigh to myself. As is my morning ritual, on the way to the bathroom I flip open my laptop. It's been "standing by" on my desk through the night. On the second pass by my desk, headed to the kitchen to make coffee, I make sure my browser is open and loading my default start page: the *Toronto Star* newspaper. I grind some beans and load the coffeemaker in the kitchen, return to my desk, and start going through the headlines methodically. I click the "international" link first, scrolling through and opening each story of interest to me in a new tab. Among other things, the international news includes a story on the Iraq war that includes the civilian death toll, a story about the Palestinian/Israeli conflict that makes no mention of the United States (nor Canada for that matter), a report on a Dutch ceremony honoring fallen WWII Canadian soldiers that I don't choose, and a coverage of a freedom of the press demonstration that occurred in Paris. I hear my coffeemaker winding down its task with a few last desperate gurgles and get up to serve myself an oversized mug of my morning addiction. Back at my desk, I click the "national" link, then click the next tab to begin reading through my morning newspaper. When I've exhausted my selection of international stories, I go back to the "national" tab and scroll through the headlines, selecting a batch of "Canadian" stories: NDP leader Jack Layton has made a deal to prop up Liberal prime minister Paul Martin, Conservative Stephen Harper isn't as popular as he'd hoped to become but still wants to bring down the Liberal minority anyway, Ottawa seeks voluntary sperm donations, and the Gomery inquiry over government graft continues.

I drink my coffee, vacuously reading through the stories, my two habits combining to complete my morning ritual. After this round, I click the "Ontario" link, then get up to pour myself a refill. Resuming my seat, I see that Ontario premier McGuinty is calling out the PM over federal transfer payments, a new provincial budget is coming next week, the province is considering a couple of new nuclear power plants, and a lobby for more health facilities for the disabled is making some noise. I feel some satisfaction in this last point, hoping that somehow my brother can benefit from any improvements that way. I silently think about calling my brother for a moment, then look back at the screen. Finally, my mind is starting to rev up a bit from my caffeine fix, I click the Greater Toronto Area (or GTA) news. Here I read that the Crown Prosecutor wants a certain pedophile jailed for life, a school bus driver was fired for drinking on the job, a British anti-Semitism incident is having a fallout for the University of Toronto, and not enough is being done to promote cycling as an alternative to cars in downtown Toronto. I quickly look at *The New York Times* in a similar fashion, but since there are fewer headlines and most of them have something to do with either aggrandizing the United States or pointing out how the government has screwed something up, I actually read through very few articles. Such content holds little appeal for me. I look at the time. 9:45 am. I grumble to myself: "Damn...better get ready to go to the office..." After selecting my wardrobe, I walk outside into the sunny but cold Illinois morning and strike out in the direction of my desk at the university where I know a stack of journal article submissions await processing.

* * * * *

Morley suggests that media acts as a conduit connecting the interior of the home to the outside world (2000, p. 87)—in this case, the driver's license stabilizing a connection between myself and my home city. In an actor-network understanding of this relationship, such a conduit (at the behest of the user) acts as the "sticky stuff" that holds users and the world they are navigating through in a sufficient state of tension to stabilize the actor's autonomous translation of their experiences. Mediation locates home, allowing one to consume it as both content and as a ritualized performative script. This upsets a commonly held assumption about the relationship between distance and power. Massey writes that "drawing of boundaries is an exercise of power" (1994, p. 69), the idea here being to keep at a distance that which is considered a threat to the comfort of the home. However, what is important for feelings of comfort and safety is not so much the actual disconnection from the outside world (which in practice seems rather impossible to achieve), but rather the constant maintenance of the *aesthetic* of an autonomous construct of home as an entity—in ANT terms, maintaining home as a singularity. Power is an effect of a network of alliances that allocate aesthetic distance between self and other. The very personal motivation of such alliances is the production of "ontological security."

"Ontological security" (Giddens, 1991) is an emotional sense of security and confidence in oneself that develops from having active, trusting relationships with others, traditionally established in the family and maintained throughout daily life. Giddens describes ontological security as "confidence or trust that the natural and social worlds are as they appear to be, including the basic existential parameters of self and social identity" (1986, p. 375). He explains that the modernist propensity of holding danger apart from oneself, keeping it untouchable and, therefore, unconfrontable/unmanageable, results in the seeking of safety through routinization of the everyday (Giddens, 1990). He describes: "Ordinary day-to-day life...involves an *ontological security* expressing an *autonomy of bodily control* within *predictable routines*" (Giddens, 1986, p. 50). This concept is rooted in the idea of personal practices of methods that control anxiety, a sense of trust built and maintained through everyday routinization, or perhaps more directly stated, trust through predictability of the behavior of other humans and the natural world. In my reading, this routinized stability is the very thing ruptured by the technologically induced time/space dilemma theorized by Virilio (2000), Baudrillard (1988), Burgelman (2000), and Fortunatti (2002). Thus the motivation for routinization in social conventions is the construction and maintenance of security. An actor-network analysis of this concept would have to consider this a project of stabilization, the practice of autonomy with the goal of maintaining one's singularity. In routinization, action becomes machination, attending to the function of the network as an apparatus. Mastering mediation works to ensure one's dominance, stabilizing alliances to singularize oneself as a predictable identity in this context: maintaining an alliance of media works as a routine practice of self.

Technological Nationalism at Aroma

This could be any time, at any hotspot in the world. But it's not. It is, in fact, a typical Tuesday evening in Aroma Café, downtown Champaign, Illinois. Every table has one or two students with laptops open, and about half are wearing headphones. I put on my own headphones, open my iTunes player, then look for the document I wish to work on. As I pull together my notes to write this very vignette, I am listening to a streaming broadcast of WeFunk radio, a Montreal station that I believe is the best urban station I've heard. They're not afraid to be raw sometimes. I hear all about the club scene in Montreal, the mystique created by WeFunk's awesome grooves attesting to the magical hipness of the city. I listen to the DJ uttering in a distinctly Canadian accent a distinctly Canadian expression: "Indeed...Indeed..." Who in the world but Canadians would think "indeed" is a cool enough expression to mix with HipHop? Or that calling someone "buddy" makes you sound tough? It makes me feel like Montreal is a place I want to be. How come it's easier for me to feel connected to Montreal from a little coffee shop in Champaign, Illinois, or from my brother's living room in Toronto, or from a hotel room in Vancouver or Tokyo or Beijing, than when I'm actually there? I don't know.

* * * * *

One contemporary facet of ontological security is a construct of alliances through distance. Mediated rather than face-to-face communication has become the basis of many relationships. Communications technology is used in the postmodern city as a way of securing the space between self and other, "home and not home" (Spigel, 2001, p. 401). However, home is always entangled with travel and a notion of not-home (Morley, 2000). In these configurations, there is a Derridian reliance on "other" to hold oneself stabilized within the hazard-loaded networks of everyday life. Every relationship represents a bridge between self and the "other," or here and there—a supportive dependency on each end. The world becomes de-distanced through the normalization of a "uniform distancelessness" (Fry, 1993; Heidegger, 2001). Ubiquity of communications technology is not enough in itself to stabilize feelings of safety and comfort. Just as owning a house does not ensure having a home, it's not just having the technology that achieves network stability; network stability depends on how and when and with whom technology is used. The goal is a constancy of ontological security regardless of borders and other territorially demarcated constraints. This can be accomplished through personal devices that make some of the comfortable and secure performances of home portable.

I use the term "mobile" to describe some of the features of home as portable technologies, miniaturized but no less comfort-producing, constructive of ontological security, and no less alliances in the performance of self as an identity. In the contemporary world, borders are easily exposed as imposed fictions. Movement exposes the spectacular nature of authority, and the absurd arbitrariness of the machinations that are employed to maintain them. Guattari (in Stivale, 1985) theorized the portable nation, enacted in the practices of the territorial travels of its perpetually nomadic population. Being automobile is not just desirable, but rather it is a fundamental part of modern citizenship. There is no one ideal cosmopolitan/global citizen, but rather a multitude of citizens who express in varied ways, in different places at different times (Woodward, Skrbis, and Bean, 2008). Portability and miniaturization of technology allow one to "familiarize" personal space irrespective of location. Williams suggests that "mobile privatization"—the construction of mobility as individual activity—is a regime of privacy. Hay and Packer (2004) apply this language to suggest that automobility is a way of performing privacy, something that is usually associated with home. Like other actants, portable technologies are active in the construction of one's ethics ("one's relationship with oneself"; Gauntlett, 2002). Portable technologies are used as part of one's own individual productive apparatus, as an affirmation and a set of practices and identities within systems of power (i.e., "technologies of the self") in addition to their uses as appliances or tools for work. In relation to automobility, "technology of the self" is identifiable as "a set of technical operations and skills necessary for properly conducting oneself within a current regime of mobility, which supports a social arrangement and a moral economy" (Hay and Packer, 2004, p. 18). The actor-network concepts "inscription" and "sociologism" reflect this performative necessity, implicating the human actor—one's "self" as a performed identity—as the dominant voice in the translation of home. In sum, home is what and where one chooses to make it: sometimes a nation, sometimes a city, sometimes a family, sometimes a house, sometimes friends, sometimes strangers, and so on, ad infinitum. As Law (1992) and Latour (1986) point out, social networks defy theoretical completion, but in practice they, in fact, can be traced with a sense of finality by simply listing the relationships that constitute the appearances of singularity. Such is the case with this description of "home," and such is one important aspect of several of the ethnographic vignettes presented here—they serve as a listing of performances of home.

I speculate, additional to this elaborate theorization of my structure of feeling, that home is also a place to begin from, and a place to return to. Home is a default "us" to position against an unknown "them," a default Heideggerian "they" whose chorus of speculative wisdom stands as a reference for you and me. Every computer has a home (just look at your keyboard and see), every web site has a home page. One must have a home to get a cell phone plan. Most important for this book, home can be as mobile as oneself—an identity practiced as the maintenance and subversion of alliances, towards the stability of one's ontological security. Home is, in this instance, also the fulfillment of a theoretical foray by which I also seek to connect myself to my intellectual roots: the Toronto school of communications. I would be guilty of negligence to pretend not to fantasize about participating in the Toronto legacy of communications theory, to not (however unjustified) envision myself as part of a historical lineage of groundbreaking Canadian academics. Hence, I *must* theorize, because theory is part of the performance of being a Canadian communications scholar. But then, this is a particular type of theorizing.

It is not my intent to quantify the relations of home ad infinitum but rather, as Latour (1986) suggests, to demonstrate this theory of globalized performances of identity with some examples of everyday practice, to carry out the elaboration of a list that constitutes some of the messy networks of the world I participate in. The list of highly mundane practices that tie me to home through wireless mobility includes everything from news media consumption to online banking, family and friendship maintenance to online shopping. Anything that can be performed through an alliance with the global telecom network has the potential to become an intersection in the network's global array.

Of course, I am writing about not just myself as an idiosyncratic actor. In general, people voraciously ally these devices for their own purposes in every location possible. Given the inscribed behaviors of the devices and the sociologism involved in their enlistment, there is a great deal of phenomenologically similar performativity involved, even across cultures. I can't learn of all the personal feelings that go along with the experiences of other people I have witnessed, but I can voyeuristically recognize and record the ways that devices work as actants in everyday moments, demonstrating how we might have Heideggerian eruptions of truth even in something as simple as the act of text messaging. It is perceiving des-inscription (breaking with prescribed behavior) that often reveals the point at which one network opposes another—for example, when the culturally loaded performance of flirting and love usurps the machinations of teaching and working on a laptop computer. Wireless devices become participants in love, nationalistic resistance and hegemony, and other performances of status and culture. Nonetheless, it is undoubtedly the modern state's drive to totalize and dominate (Kien, 2004a) that has the unrelenting potential to exert its hegemony as translator of the actor-network at any given moment. Beyond reified notions of static borders containing dead territories, the performance of nationalism allows the re-enframing of the nation—the "homeland'—as a cultural territory. What does this performance look like? How can one identify the nation as cultural performance? An example can help answer these questions.

Enframing the Nation as Cultural Territory

I'm watching a rather bad movie on TV, the title of which I don't know, subtitled in Korean. The story begins with a bomb blast in Montreal and then moves to a scene in Washington, DC. I'm fascinated by one scene in particular: in order to get inside a U.S. senator's office, a U.S. agent decides to disguise himself as a Canadian government official presenting the senator an award for his bravery in the face of the Montreal crisis. However, it is the method of cloaking that catches my interest. The protagonist and his partner camouflage themselves mainly in dialog—Canadian vernacular and Ottawa Valley accents—and the performance of excessive "Canadian-style" politeness. As the door closes behind the undercover agent, the senator remarks, "God... did you ever hear so many thank-yous?" This might have just passed me unnoticed as a bad American caricature of Canadianness, but for two vital pieces of information. First, it is none other than "famous only in Canada" actor Alan Thicke playing the part of the American senator, and second, my experience of living in the United States has taught me very well that most American's tend not to give anyone else's behavior a second thought; I simply can't imagine a U.S. senator devoting a single second to comment on a foreigner's manners. The scene is a hyper-Canadian imagining of a hypothesized American caricature of Canadianness.

As I shut the TV off to head out for the evening, I find myself perplexed at this strange intersection of performativity. If I hadn't come to Korea, I'm sure I would never have seen this second-rate movie clip (I certainly wouldn't have bothered to watch it in Canada. And yet there it was, a Canadian playing an American (no surprise—somewhat a Canadian thespian tradition), but with another actor performing Canada for him. I think of some of my own experiences performing Canadianness in the United States...my unease with what seems to me to be the brash militaristic mannerisms and love of conflict in even the most "docile" of Americans, my initial perception of what I condescendingly interpreted as Americans' childish inward fixation on themselves, the lack of any notion of social organicism and their disdain for any deference to others, the overt racism...in a truly Canadian performance of identity, I assemble a definition of myself based on trying not to perform the stereotypes of Americans that I find repulsive, ironically trying to perform the ideal American citizen, just as Canada as a nation often seems to try to show the United States what it should be instead of just being whoever "we" are. The film depends on Washington, DC as an opportunity to portray hyper-Canadianness, and I in turn depend upon my own experiences in the United States to understand that it's not the one actor's lines or performance that makes the scene Canadian. Rather, it's the fact that it was felt necessary for that absurd little scene to exist at all that nationalizes Canada as a cultural territory through the movie.

* * * * *

These are the markers of nationalist performativity: nationalist chauvinism and condescention to the "others," caricatures constructed through hyper-performativity, familiarity with the hegemon such that belligerence is felt possible, the felt need to insert the signifiers of the nation into otherwise mundane interactions and nondescript places, and the seeking out and recognition of familiar cultural scripts. Nationalism has a tendency to dominate all other identity discourses (see Gilroy, 2001), both enframing and enlisting discourses of race, gender, ability, and other identity-based definitions (in the Canadian mind-set, "our" people are always better looked after than "their" people). Through strategic territorialization, the nation also usurps the house as a place of residence. Although at times performances of love and friendship refocus one's attention away from the nation as the ultimate token of power, nationalism always lurks in the background, ready to be called into alliance. If the global telecom network is the primary physical network of wireless mobility, nationalism has found a reliable ally in its extensions of cultural territory.

Maurice Charland's deconstruction of the Canadian nation proposes two territories of articulation: the physical terrain and the cultural territory. Already the territorial parallels with the physicality of the telecom network and the network content as a cultural territory are apparent. Charland tells us that the rhetoric of "technological nationalism" argues that the Canadian nation is constituted by space-binding technologies (1986, p. 196). The physical territory of the nation was brought together and maintained by the railway (and later, highway system), binding physically through transport. However, this situates technology "at the centre of the Canadian imagination, for it provides the condition of possibility for a Canadian mind" (1986, p. 201), hence the term "technological nationalism."

In contrast, the cultural territory of Canada was given shape with the radio and other electronic media that bind culturally and ideologically. Electronic media provide "the site for Canada's cultural construction" (1986, p. 201). Throughout Canadian history, the threat of U.S. media expansion warranted attention from the Canadian government to bind and thereby secure Canada as an imaginable and imagined nation. The Canadian Broadcasting Corporation (CBC) is the main instrument of the state in this process, which has been exceptionally effective. If Charland is correct, Baudrillard's "neo-tribalism" is not new for Canadians but is rather all we've ever really had. In the context of globalization, the separation—perhaps, even bifurcation—of these distinct territories is not just a Canadian phenomenon: "all over the world, the processes of migration, electronic mediation and the work of the imagination of the great masses of the people have effected the separation of culture from place" (McCarthy et al., 2003, p. 13). This is exactly the physical/cultural spatial split that Charland theorizes. Working with Charland's premise, electronic mediation is inseparable from nationalism and personal identification. For example, the CN Tower is not only mythological and spectacular in being Toronto's most unmistakable icon and world's tallest free-standing structure, it also happens to be an important, functioning communications technology.

IMPE, CBC, UIUC CC

It's a typical Wednesday evening at the onset of spring. I'm feeling good after my workout, on my way to the change room at the Intramural and Physical Education building (aka IMPE) in Champaign. Less than 10 meters from my destination, I'm stopped in the hallway by Michael, a Montrealler who I've spoken with a few times in the gym. We've spoken before several times, so I know he's an Urban Planning MA student. This evening he is somewhat animated, seemingly eager to talk. I hesitate a little bit, since the fact is I really want to be somewhere else, but my Canadian manners won't let me walk away. "How you doing…?" he asks.

"Good…How are you?" I reciprocate. We chat a little bit. He tells me that next Wednesday he's decided to hold a meeting to start a Canadian student association on campus. Knowing that I don't have much time, he tells me in a rush that the club will meet every Wednesday, and he wants to have all kinds of Canadian cultural events. He wants to have Canadian movie nights, organize pub nights that feature Canadian beer, have Canadian food events, and advocate for the local cable TV company to carry CBC Newsworld…

"What?…Wait a second…," I stop him. "Really? You want to advocate for the cable company to carry CBC?"

"Yeah," he affirms, "they do it for Spanish programming, so why not for Canadians?" I fall silent for a moment, letting the significance of what he's saying to sink in.

"Yeah, sure...," I say. "You know, I watch Newsworld and the National online whenever I want...," I tell him. "Even *Canada Now* is online...Did you know you can stream them on your computer?" I ask.

"No," he says, pausing for a moment. "But I'd like to see it on my television anyway, and some French programming too," he tells me. I think about it a moment.

"Yeah, sure, sounds good, I'm glad you're getting an association together," I say encouragingly. We discuss a couple of other things and agree that I'll watch for his email.

A couple of days later, Michael's email arrives. As part of my morning ritual, I read through the content, most of which I already know. I scroll through to the bottom where he has included as "advocacy projects" getting the following added to the local cable selection: CBC Newsworld, TV5 (French news), an extra-charge Canadian package of channels, a French music video show, and, an almost throw away item, getting Canadian food in restaurants and in grocery stores. "Wow...," I exclaim, "Seems like he wants to watch a lot of TV...This is like major media dependency here...Is that the Canadian way?" I wonder. I close his email and turn my attention back to reading the *Toronto Star* online.

* * * * *

In the contemporary context, network has come to metaphorically characterize all human space: "There is no longer a place that can be recognized as outside" (Hardt and Negri, 2000, p. 211). Subjectivity is constructed in networks: "no subjectivity is outside, and all places have been subsumed in a general 'non-place'" (2000, p. 253). The nation as cultural territory is, through mediation, a culturally spatialized network. However, Cvetkovich and Kellner (1997) point out that macro-perspectives such as this, accurate in some ways as they might be, tend to present totalizing and reductionist accounts of globalization as a singular phenomenon. Analysis of localities and particularities of everyday life present a more accurate portrayal of globalization as it is actually experienced. This allows us to "see how local forces and situations mediate the global, inflecting global forces to diverse ends and conditions and producing unique configurations for thought and action in the contemporary world" (p.2).

National identity persists, expanding in reach in spite of the rhetoric of "borderless" globalization. The performance of mobility "sustains both conscious and unreflexive impressions of national belonging" (Edensor, 2004, p. 111). "Embodied codes...guide actions in particular settings" (2004, p. 112). One need look no further than mundane everyday habits for ethnographic demonstration of this phenomenon. An Internet browser start page becomes an exercise in technological nationalism when it is set to one's home newspaper and, of course, the laptop can function as a broadcast medium, keeping concerns and vernacular from "back home" at the fore of one's thoughts, regardless of one's dislocation. The spatialization of national cultural territory may thus accompany one throughout the globe, but equally within one's physical home territory. The result is a consistency in the maintenance of a national imaginary that can be relied on throughout the globe, regardless of location. Wireless networked devices help construct the "self" as a national identity no matter where one is and acts as a cultural artifact in everyday rituals despite physical displacement, assuring ontological security.

CHAPTER EIGHT

Technology Is Human

Things Break Down

"Damn it! What's happening!!??"

I look at the blank screen angrily. I try restarting my laptop. No luck. Then I move the screen. Suddenly it flickers. Then, it comes on. Then it dies again. I move the screen again, and it lights up dimly. I move my whole computer, and a strange pattern of digital noise appears on the screen. It goes blank again. I determine my computer has developed an intermittent screen malfunction. I'm now afraid to move it at all, in case I cause it to completely blank out. But, of course, I must at some point. Choosing to take it to the shop on a Friday would be pointless, since I already know nobody would look at it until Monday. So I keep it open, on, and try not to move it or bump the desk.

After a weekend of precarious surfing, Monday comes around. After performing my morning routine as well as possible, I carry my little ally to the repair shop. At the repair shop, I talk to Jacqueline, the Mac expert. She and I are well acquainted by now, since I have been in to see her numerous times already. I've already replaced the motherboard, the hard drive, and then the motherboard again. Man, I'm hard on computers! As I expect, she tells me it will be at least two days before I could expect it back. As she has in the past, she tells me a story about the recent problems she's been encountering with Mac computers. As I have in the past, I silently notice the scars on her wrists again and look into her eyes wondering silently about the pain that caused them. I thank her, and I expect that I will have to be without my computer for at least a week waiting for parts after her initial inspection. I already know for that period of time I won't be able to write or record or work with photos or web sites from home or in the coffee shops where I normally spend half of my writing time.

A week later I call to check up on my computer as I've done in the past and hear that Jacqueline has suddenly decided to take a day or two off, as I've heard before when I've called about past repair jobs. I wonder for a moment what malady, new or old, this incredibly tech-savvy woman has been burdened to contend with. I readjust my everyday life to work on site at the university, getting my fill of Canadian media content through alternative apparatuses. Far from being the interface of some sort of emancipatory apparatus, my computer becomes yet another mundane problem in the embodied everyday life that I live. More importantly, in spite of my computer problems, I hope most of all that my favorite computer technician will be there, whether I need her or not.

* * * * *

Theorizing cannot reveal the very personal emotions that I experience as rooted in intimate interpersonal relationships rather than in technological systems. In de Certeau's (1984) paradigm of the everyday, lives, spaces, and organizations are planned strategically, but we experience the everyday tactically/temporally. As Urry comments on mobility in the everyday with Heidegger in mind: "People only go through spaces in ways which sustain them through the relationships which are established 'with near and remote locales and things'" (Urry, 2000, p. 132). For Heidegger, time is used to structure the everyday, poesis is revealed in disobeyant moments, and, therefore, to find truth, one must seek a subjectivity that expresses tactical agency in everyday life, a subject that is ordered but isn't compelled to obey, a subject that is possible to find in every level of consumerist/capitalist hierarchies that demonstrates part of the body still belongs to the earth.

Virno (2004) theorizes that the mobile worker, unable to identify any solidity in their sociopolitical existence, seeks security by already existing in the routinized state of what Hardt and Negri (2000) call the "being against." The technologically mobile subject is "dwelling-in-travel" (Clifford in Urry, 2000, p. 133). Perhaps some promise is held in advancing the methodological frontier of technography from a qualitative approach, seeking to understand how we experience the technologies we surround ourselves with as actors in our social networks. What is more obvious is that the question in the present moment of technographic social research is not what is old or new, modern or postmodern. Rather, the task thus far ignored is to reveal what continuous practices are performed and represented as new in spite of their sameness, and where fissures present the possibility of experiencing authenticity. What is gained or lost in this re-representation of the old as new is important, especially so if investigating a hypothesis of ritual culture (Carey, 1989). In our modern Western civilization, one might suspect that what continues to be prized above all is an emphasis on speed and efficiency, but perhaps where other cultural values are at play, a different emphasis—perhaps, something like sociability—might be revealed. In the moments where cultural actors figure out how to reenact familiar scripts in performance with new artifacts and/or in the presence of "other" cultural values, rupture in the dominant enframing is temporarily possible. The question from this perspective is not whether technology signifies freedom or entrapment, but rather, how people experience technology in agreement with or in spite of the disciplinary regimes and social conventions built into them. Although hopeful of avoiding the entrapment of an unquestioningly reproduced enframing, such an approach also potentially lays bare the problems of romanticizing mobility as the "new normal" of an overly glorified emancipatory hope, since technological mobility comes with it's own set of discriminations (i.e., ethnic chauvinisms and the same range of "isms" we know as performative social ills, and economic issues such as the digital divide).

Computer Love (Scene 1)

The brand new HP laptop sits open on the table top, while the closed Dell sits forgotten across from it, napkins tossed casually onto it. It's the last day of February in Champaign, Illinois—a cold night. She is a middle-aged brunette woman I often see in the coffee shops where I have done a goodly portion of my writing. She is often alone, with the slightly brooding countenance I am used to seeing performed by women who I suspect feel "too old to be single." He is a pudgy middle-aged man wearing a tucked-in blue tennis shirt, ample roll escaping around the braided leather belt line of his khaki pants. I've never seen him before. He projects the image of someone who "does his job." His chair is parked in the aisle beside her. His blue eyes dart quickly to mine, then away. She laughs as he points to something on her screen. He is serious. He explains something technical to her. She earnestly carries out his instructions and then thoughtfully asks a question. He performs his role of male/expert. She flirtatiously attends to being his student. Her laughter animates the atmosphere of the area we share. His serious eyes meet mine once again, quickly averted back to her. Does he see me as competition?

He looks thoughtful for a moment. He gets up to refill his cup. She waits a few moments and then jumps up and follows him to the line, happily asking a question. She sits down again, and when he returns, he seriously hovers over her left shoulder, watching as she tries something else. He explains something to her without smiling, then he sits down across from her, his back to me. Placing his coffee beside the Dell, he finally opens the cover. From over his right shoulder, I see that she flirtatiously asks him something else, continuing her lively chatter. I hear him chuckle deeply, his shoulders jumping up and down. He takes a sip of his coffee. She points at her screen and looks thoughtfully at it as she states something in a serious tone. She rests her pointing arm on the table and then touches her face with the other. She taps her lips with the red ballpoint pen that has materialized between her fingers. He coughs, too loudly. She packs up her laptop, sliding the computer into the black backpack on the floor behind her chair. She carefully winds up the power cable, tucks it in with the laptop, and removes some papers. She looks down at the papers she has placed in front of her. The back of his head looks like he is doing something on his computer.

She stands up, walks around the tables as if exercising, stands in front of him and makes a strongman pose. I see the back of his head turn to look up at her. I imagine him smiling. She sits down again. They work for a while, diligent, attentive to what is in front of them.

A quarter hour later, their conversation is renewed. Suddenly, she springs from her chair and scoots over to stand beside him, looking at his screen…she is leaning over, pointing at it, animated, talking…I see the back of his head moving up and down, back and forth…she is now on her knees beside him, pointing at his screen, talking, gesticulating…she is back on her feet, pointing, talking, his head bobbing…she gently returns to her seat, face now presented earnestly to him, talking seriously but intimately…suddenly, eyes narrow as a smile erupts, his shoulders and head bobbing up and down with laughter…

Quietly now, but still with a half smile, more talk from her…more bobbing up and down from him, his arms now crossed and placed on the table in front of him. Their laptops are now ignored completely. This is "real" face time. Serious. He turns sideways in his chair, presenting her with his left shoulder. He surveys the room quickly, seriously. Her chatter continues without pause.

He slowly puts his laptop in the brown shoulder bag sitting by his feet. He stands to leave. His pants have holes worn in the back pockets. He puts on a khaki fleece and then pulls on his dark blue overcoat. Her voice gets louder. She gestures broadly. He laughs loudly and then bends over to pick up his leather shoulder bag. She watches him brush past her towards the door. As he steps into the cold night, her face is suddenly serious and tired. She shuffles some papers in front of her. She puts down her pen. Lips pursed, she jumps up from her chair and walks right past me, disappearing into the washroom.

* * * * *

Heidegger tells us freedom is achievable only technologically, by understanding the essence of technology as the means to understanding Being, and by understanding technology as an intimate facet of being human. Humanity is the engine of technology, but humanity is in danger of being trapped in the endless paradoxical/cyclical reproduction of the dominant ordering (enframing), looking only inward at the peril of ignoring the awesomeness of the truth of the animate universe. Modern technology enframes the world in such a way as to create "uniform distancelessness" by embedding the original distance in our media, preventing us from gathering the universe to us, and preempting the potential for authentic experience. Distance is needed in order to conceptually set things apart, and thus enabling things to be gathered, to bring together the elements necessary for the creation of things and thereby to set truth into motion to reveal itself. The end of distance thus effectively keeps us from discovering truth. We're always already predisposed to ordering humanity into the standing reserve along with the natural universe in general, as having only use-value in a schema of technological ordering rather than existing independently. Humanity is thus ordered in such a way that the body and being become a conflated worldly concept, kept at a distance in the standing reserve.

One might say Heidegger leads us to an intellectual space that is both dismal and hopeful. There can be no hope of an en masse liberation of humanity, but at the same time, individuals can find truth if the essence of one's existence in the universe can be adequately brought to experience. However, I propose that contemporary mobile technology (in contrast with but not radically different from more stationary appliances) presents the individual subject with an intensification of the modernist technological challenge, a challenge that is all the more formidable because of its effect of profoundly embedding the modernist structure of everyday life even while creating the illusion of deconstruction through personal mobility. In explanation, at risk of romanticizing Western society in the past as somehow more spiritual than the present (a type of thinking that I affirm I *do not* want to promote), I turn to Durkheim and the concepts of sacred and profane to represent the dominant enframing of Western spiritual practice. I suggest, with Heidegger's explanation of ontology in mind, as distancelessness is unable to conceptually isolate things, and from a position in the standing-reserve, that the technologically mobile subject is unable to be sure of what is sacred, what is profane, to be sure there is any such thing as truth, to know if these are even useful concepts any more, and that even if they are, that the embedding of distance works to prevent the transformation (gathering) of one into the other in order to experience truth.

One need not have faith in technology, for looking back tells us that technology inevitably breaks down and fails. It has no motivation to get back up. A crashed system does what it is programmed to do in such an instance: nothing. Rather, faith in ourselves as motivated actors can drive us onward. We fix things when they break. We strive. When human beings stumble and fall, others may pick them up, carry them a while, and then, when they are ready, set them on their own feet again. Together we fix things that get broken. What causes this teamwork is unimportant here; it is a social fact that it happens, and it is the effect that gives meaning to our actions. Our networks (and make no mistake, every network is social) might often be conflicted, and we as actors might often be struggling, but what might be overlooked is that many of us win with incredible frequency. I venture to say that small, mundane, unmentioned accomplishments in everyday life comprise most of the human experiences I have taken the opportunity to look at. But, of course, a victory for one is not necessarily a victory for all. And this is, of course, our utopian ideal and ethical imperative, however impossible it might prove to make manifest in the "real" world. How can we use technology towards such a lofty goal?

The vignettes throughout this book show that the effects of technology are dialectically connected with everyday performance, and imbricated in various messy networks all at the same time. Which one gets paid attention to depends on which translative vocabulary is dominant at that moment. It is thus not just the programmatic machinations of technology, but technological performance that we can turn to for the fulfillment of the ethical ideal through the fruition of technological achievement.

Computer Love (Scene 2: Jin Hu Cha Can Ting Restaurant, Beijing)

It is lunchtime at the Jin Hu Cha Cantonese restaurant in the China Business District. Although there is wireless Internet here, for once this isn't the reason for my choice of location. Rather, it is across the street from the business center where I have just faxed some documents. The cool, dry air conditioning is a welcome contrast to the hot, muggy dust that envelopes Beijing on this day. I don't have my own laptop with me, as the technology didn't figure in my afternoon plan. However, I take the opportunity to look around at a few other customers with open laptops. One table in particular catches my eye: a young woman and a man sitting side by side in a booth facing my direction.

The woman's laptop yawns widely from the table towards them. The man's computer sits closed on the opposite side of the table, suggesting that they, in fact, started out on opposite sides of the booth. They are both attired in what I understand is the Chinese business casual uniform. As I observe them from behind my lunch, I see her using her hands on the keyboard while they converse in what appears from their countenance to be quiet tones. He keeps his hands on the table in front of him. Although they maintain a "respectable" visible distance between their bodies, I can't help suspecting there is more here than meets the eye. Patient observation proves me right. They both occasionally gesture towards the screen, sometimes at the same time, sometimes suspensefully bringing their hands suspiciously close together. The laptop is now an ally in their love. In what I recognize as the female performance of stereotypical Chinese flirting, she keeps her eyes forward, fixed on the screen, maintaining an artificially stoic aesthetic of disinterest that, in fact, signals exactly the opposite. He looks at her, jokes, smiles, and touches her shoulder happily, all the while maintaining the careful distance between them. Smiling in turn, she directs his attention to the screen without looking at him, now pointing, now laughing, now serious, now nodding...Her performance shows that she is very conscious that he is closely watching her. They continue like this for the duration of my lunch. In the name of science, I am a careful voyeur, mentally noting every nuance of this carefully enacted performance. I think back to an evening at a coffee shop in Champaign, Illinois, and smile to myself about how fundamentally the same we human beings seem to be, in spite of locations, in spite of technologies, in spite of the enormity of the myths we use to justify our actions. In the tawdry details of our everyday routines—having lunch or coffee, for example—the magic of human love suddenly erupts. Forty-five minutes later, I get up and pay the bill. A last look over my shoulder as I leave finalizes my assessment: she has sat back in her seat, put her hands in her lap, and is now looking directly at him and laughing; he, staring into her eyes, with animated gestures tells her something that unlocks her stock of happiness. The laptop, now forgotten, continues to sit open as silent witness to their lunchtime affair.

* * * * *

There is a nagging propensity to characterize the advancement of technology in terms of cyborgist hyperbole and posthuman alarmism or its extreme opposite, euphoria. Although such theorizing may describe some dimensions of the human/technology relationship quite well, it has been my goal to show that in spite of technological potentialities, life everywhere is lived "everyday," be it in Canada, the United States, Japan, South Korea, China, or the mountains of Mexico. Everyday experience thus enframes the technological experiences we participate in. Translated through desired outcomes, technology is both good and bad at the same time, tactical and strategic, but always instrumental. Despite the beneficent design intentions of inventions such as those showcased in Bruce Ma's roaming *Massive Change* exhibit, technologies get used in all sorts of ways that their designers couldn't possibly imagine and sometimes don't get used at all for the very same reasons. Technology impacts through both its design and its use, and through whatever else happens in the lead up, duration, and thereafter in the culturally informed performances technology participates in. It is no secret that people don't always do what they *should* do, nor does the word *shouldn't* always have the desired preventative impact.

Yes, as Heidegger aptly pointed out, humanity and technology may be inseparable. However, I cannot, as some do, interpret this as a hopeless situation; pessimism is a luxury few in this world can afford. As did happiness and joy, danger and suffering most certainly preexisted our present technological situation. Wireless technology can do very little to improve the circumstances in which malnourished refugee children in a remote southern Mexican mountain village live as victims of the human motivation of unabashed greed. But sometimes, when used with cunning, when the circumstances align just so, even very little technology can be enough for a short time. Even if the United Nations does, in fact, count the presence of television as an indicator of the quality of life, not everyone likes TV, and not everyone watches it the same way. And so, unabashedly pedantic, I say to pessimists, alarmists, optimists, social determinists, and technocrats alike, let's not give up but rather take heart and have faith in what makes us human; let us *be* the engines we are, revealing life for what it is in every moment, including this present one. Technology doesn't bring us closer to one another—that is not the work it does. But somehow we do sometimes get closer together, in spite of the normalized distancelessness of globalization. Even if it doesn't bring us closer to each other as people, technology is the only thing that can bring us closer to the truth. The future is undoubtedly as technological as the past, but we may go bravely there knowing that our mobility, being, and global network are as human as the dreams that give rise to the technological moments along the way.

APPENDIX

From Heidegger to Technography: A Way Outward in a Distanceless World

Do Guns Kill People?

The nature of the relationship of humans with technology is a topic of relentless discussion. The circumstances we find ourselves situated in—our contemporary "hi-tech" civilization—may in some individuals tend to emphasize a feeling of servitude to the technological systems and devices that we employ to negotiate and maintain our daily lives. Conversely, a sense of euphoria induced through mastery of hi-tech devices may give some people cause for celebration: what could be more exciting than the feeling of being truly in command of one's own destiny? I suggest there is more likely to be some of each orientation in human subjects rather than a caricatured extreme of either, but more intriguing for me is the silently assumed, undefined, but nonetheless very present idea of "freedom" equally expressed by both voices in this feigned dichotomy. Perhaps this obsession with an unexplored idea of freedom and the fear of being entrapped by humans or machines alike is a reflection of the psychosis induced by the mantra of "freedom" chanted throughout Western civilization since at least the French revolution, and by none so vehemently as the citizens of the United States. Although there is much public debate about freedom in terms of civil liberty and political systems, when it comes to the issue of freedom in relation to technology, the legacy of the Luddites and other rebels against automation seemingly continues to dominate the popular mindset of a paranoid public. Popular movies such as *The Matrix* and *Terminator* trilogies, the classic *2001: A Space Odyssey*, the trend-setting cyber movie *Tron*, the timeless *Bladerunner*, and the *Robocop* movie serial (to name just a very few) reflect both celebratory and doomsday attitudes regarding humanity's tumultuous psychological relationships with the technologies we create and depend on, though one might be hard pressed to find examples of any actual battles exclusively between humans and machines. The Luddites smashed their machines very well, but I dare ask how many Luddites were smashed back by the machines in the "battle"?

The question regarding the nature of the human/technology relationship is often summed up in one challenging either/or question: Do guns kill people or do people kill people? Though provocative, I judge this a false dichotomy. Both guns and people kill people. As Heidegger might say, the question obscures the truth. Killing was not necessarily invented by people, is not owned by people, and is not the exclusive domain of humans. And guns don't kill just people. Although technology may kill or cause accidents, killing is neither an accident nor an addendum of technology. Death *is*. Accident and intention adjudicate the time and place of death—the killing—not the desirability of death itself. At the very least, we can be fairly certain of the deaths of entities other than oneself, but we can positively identify neither death's invention nor its demise. We know killing only through the work it does—to bring about death. Death is the revelation of obscuring, the demonstration of paradox, horrific in its immortal grandeur, and inhumanly perfect in being a flawless finality of all worldly ends. The dead are consequential to the work of death only in that they allow it to be revealed. The essence of life is another such thing, denying authorship other than reference to another "original" being other than one's own, be it a deity, a parent, science, or what have you. Guns and human behavior are tools towards the final work in and of itself.

What is human? Again the question hides the answer. One is tempted by another false dichotomy: try, for example, the "evolutionary" struggle of man versus nature, or perhaps the cybernetic conundrum of life versus machine. I suggest that to be human is to not be apart from these entities and struggles, but to be profoundly bound to them. The overwrought cybernetic postulation of "living" machines (Wiener, 1954, p. 33) begs the question absent from debate: can machines be dead? If so, should we have funerals for our spent ballpoint pens and discharged batteries? Death reveals the goal of life (or at least one goal of life) as the fulfillment of a "life" program, the evolution from physical selfhood to...whatever (spirit, absence, oblivion, divinity, eternity, reward, damnation, mediocrity...insert your destination here). Is this struggling part of a *machine's* program? I suggest the answer is yes; however, only in so far as machines are created for and continue to serve our very human motivations. Those motivations themselves remain the appendages of humanity, not machines, and thus the fact that we do *not* hold funerals for our technological gadgets but rather reserve them for humans and dear pets.

In response to the notion that somehow our technological appendages make us "more" or "less" human (what I interpret as one idea described by Hamelink, 2000), I suggest that neither is defining the human simply a question of whether or not an animal or machine can be anthropomorphized and discussed as if they *are* human rather than as simply enacting programs that mimic human activity. To be human in our popular understanding of humanity entails something more than the aesthetics of our behavior. It is a cultural premise/norm that, regardless of the difficulty of identifying and defining it, humans have something we call a "spirit," and that it is a key aspect of being human. Popular movies—such as *Robocop* and Japanese animations such as *Ghost in the Shell* (in which cyborgs are created through body replacement/mechanical enhancement) or *Akira* (in which the "human" is assimilated into a monster by mutation)—take as their premise that in spite of cybernetic or bionic modifications, humans have souls that are redeemable. Again, I propose that to be human is to be complexly interconnected to all the material and nonmaterial aspects of being that give birth to, sustain, and terminate one's life as a human being.

What then do I consider a useful line of work for a would-be theorist of technology such as myself? My task is a very personal choice, rooted in my own messy relationship with applied digital technology and in my background in cultural and social theory. My tack is to explore the issue of freedom in relation to technology through a focus on agency, taking a cultural approach (with Carey's ritual model of communications in mind) that I ground in Heidegger's ontology. Apart from the inescapable personalization of this theme, I proceed with the hope of finding and/or demonstrating a way that most people in the world can be active in the construction of their own existence. I believe the work of Heidegger provides a philosophical base from which to develop a methodology of investigating technological social behavior that will reveal opportune moments whereby potentially anyone—not just a vanguard of hi-tech experts—can achieve some demonstrable, cognizant manifestation as an agent in their own lived technological experience. Finally, I believe this Heideggerian philosophical base contributes to methodological advances that incorporate technology as an active participant in the social actor-network—the evolution of technography as a qualitative communications research method—to more accurately illustrate the human/technology experience.

Agency in the Tele-Present Situation

My initial interest in this project stemmed from my own concerns about freedom and technology. Perhaps guilty of indulging my own romanticism, it has been my hope to help develop a program of freedom that will provide some emancipatory hope for the non-elites in this world. I find that the present state of theorizing on behalf of liberty is as guilty of overexcited fantasies of technological utopianism as any modernist since Liebniz proclaimed it would be possible to build a machine that would "free" good men from having to think "mundane" thoughts at all (i.e., how to feed humanity and mete out justice). Cyborgism (Haraway, 1999; Mann and Niedzviecki, 2001), hacking (Gunkle, 2001), Hardt and Negri's "Being Against" (2000), and other theorized technologically astute agents offer little to the billions of "real" people who, in their everyday lives, rely on much humbler forms of agency to forge their daily existence.

My instinctual response to this theoretical elitism was to look at accidental and incidental uses of technology in everyday life as examples of human ingenuity in spite of hi-tech programs of use. Although I believe this to be along the right track for my purposes, I soon found this alone was not sufficient to satisfy my own criteria, as it also falls victim to the same traps as other people's works that I'm not satisfied with. This approach again risks tying freedom to access and affluence and doesn't necessarily investigate the human/technology experience itself. Another failing in relying on this method alone, according to Virilio (2003), is the invention of the "accident" itself as a variable that is controlled for in technological design and systematic responses and, therefore, doesn't account for the social programs built into technology, the social roles technology as Latourian actants fulfill, and the social responses and inabilities to respond to technology that are part of human everyday life. Finally, relying on such a singular approach again lacks an interrogation of the notion of freedom and liberty in relation to technology, again assuming only the extreme positions that technology is either good or bad. Thus I came to think of the issue of freedom and technology as an ethical issue, needing to be examined through an ethical discourse. I found the work of Christians (1997) pointing me in the direction of Heidegger's "poiesis." Poiesis—understood as humans' participation in the creative process of revealing truth—is the essence of human agency. Thus happened the evolution of my philosophical thinking about freedom and technology, and hence the work I present here.

My main goal in this essay is to develop an outline of a theory applicable to mobile technology from Heidegger's philosophy of technology and to suggest the use of this schematic towards the development of a new qualitative approach to studying technology. I don't consider this a finished project within the parameters of what follows, and perhaps it is not even a finishable project in the longer term, given the instrumental function this schematic is meant to fulfill as a brainstorm of sorts. Rather, this crude foray into the realm of thinking is a means of bringing to hand some of the key elements of Heidegger's work in such a way as to apply them to a methodology—or perhaps choose a methodology that applies itself to his philosophy—that enables me to move forward in my own investigations of mobility and technology. The obvious question thus becomes whether or not Heidegger's own methodology—phenomenology—itself is applicable to my work.

Is phenomenology not already an appropriate methodology for studying mobility and/or technology? Many have tried to work with the ideas of Heidegger in the contemporary technological context, to greater and lesser success (e.g., Barney, 2000; Feenberg, 1991; Fry, 1993; Kroker, 2002). Although there are many complex and intriguing ideas to be found in such texts, I find that one drawback in many of the interpretations of Heidegger's work on technology is in the misunderstanding of technology as being on its own more than an instrumental equipment, and mistaking aspects of the everyday for "pure being" (da-sein) in the Heideggerian sense. In other words, it seems that some authors have fallen into the very trap that Heidegger elaborates as the dangerous aspect of contemporary technology: some have mistaken the trees for the forest, continuing to theorize humanity into the standing reserve. And although there are plenty of critiques of Heidegger's instrumental theory of technology, I have found that many authors do not see his work through to the end solution—in effect, they do not take as seriously as they could the need to elaborate what freedom actually means. Rather, many continue to assume an undefined and, therefore, mutable notion of freedom that is conveniently adaptable to their own theoretical schematic.

To be fair, Heidegger himself rather thoroughly developed the phenomenology of technology, and it is not my intention to argue with his method and his findings but to rather take up the question of what can be taken from his phenomenology to illustrate the experience of technological mobility. Of course, as a work in progress, my ruminations on the future of technology can stand only as speculations, but even so, I am compelled to suggest that the issues raised by the turn to technological mobility is an intensification of the problem Heidegger identifies as distinct in modern technology: the end of distance (Heidegger, 2001). This issue has been vigorously explored by some (Bauman, 2000a, 2000b; Castells, 2000, 2001; Held et al., 1999, Massey, 1994; Massey and Jess, 1995; Virilio, 1997, 2000) as a technologically induced crisis in Western civilization, hinging for the most part on the fracturing of essentialized identity and a located notion of self. Although this is obviously an important theoretical issue to explore in our contemporary globalizing culture, Heidegger might point out that theorizing with the intention of repairing identity and its disruption in terms of a technologically induced time/space conceptual crisis doesn't actually expose the essence or truth of technological mobility: "Such analyses of the 'situation' do not notice that they are working only according to the meaning and manner of technological dissecting" (Heidegger, 1977, p. 48). Such theorizations—just as my own work here—further re-create and maintain the phenomena of the theories themselves, the perpetuation of what we might think of as discourse. Instead of theoretically assuming that this phenomenon is destructive and must be rectified, an approach rooted in Heidegger's phenomenology might rather take such fractures as opportunities to explore and reveal the truth of being.

I suggest that what is needed in addition to elaborating how these technologies are employed in everyday life is the identification of human "poetic dwelling" (Heidegger, 2001) that takes advantage of these same crisis-inducing technologies—seizing the crisis as an opportunity to find truth and to thereby search for poetic use of these instruments where the "things" produced by our assortments of tools and actors may stand to work on their own. In terms of Carey's model of communications (1989), the task is to identify the rituals that give sacredness to the artifacts and, in the revealing of the ritual processes, stand a chance of unveiling the experience of pure being itself. The object around which this search can be organized is, of course, suggested by Heidegger himself: in art (Heidegger, 1977). My contribution to interpreting his work is in seeking to input to a method that is able to artfully represent and recreate experience, autoethnographically and dramaturgically, emphasizing the performance of culture even as I participate in writing it (â la Denzin, 2003). I believe I have found this methodology reflected in an emerging concept that, in a classic example of symmetry, I have both envisioned and stumbled across at almost the same time. It is called Technography.

Technography—which I first envisioned one evening while reflecting that qualitative communications research has failed to adequately surmise the actual experience of technology—is a new frontier in ethnography, the incorporation of technological Latourian "actants" into the ethnography of participating in the Latourian "actor network" (1992). Adopting this perspective to understand the social implications of technology, technological devices and systems must be located as social actors on equal footing with humans and thus must be treated as equal participants in so far as their codes or programs will let them in any ethnographic endeavor that seeks to study technology. I conceptualized technography as part of a broader field that I described to myself during that meditation as "Techno-Methodology." To test the originality of my thinking, I ran both of these words I thought I had invented through the Google search engine. To my surprise, I found both words already in use, albeit grounded mainly in engineering-informed approaches. My small sample of research reveals that Techno-methodology seems to entail the broader range of methods being used to research technology from all perspectives but does emphasize the need for interdisciplinarity, especially to better understand the qualitative experience of technology and the effects of culture on technology. I've encountered technography as an emerging field described as the coming together of technology, the arts, and the social sciences, exploring the inscriptions of culture on technology, and technology on culture.

I have thus set up several tasks for this essay to fulfill: first, to describe my interest in mobile technology and to then elaborate a brief description of what this technology involves and why it is of concern, and how I believe Heidegger's work is an appropriate philosophical grounding for understanding the direction in which technological development is going. Next, I will detail my own understanding of Heidegger's philosophy of technology and how it applies to the contemporary situation. Finally, I will speculate on what Heidegger himself suggests is the appropriate means of revealing truth, what I understand that to mean in terms of my own contribution to academic investigation, and further elaborate technography as an approach to qualitative communications research that will serve as a starting point towards a better developed methodology in work yet to come.

The Contemporary Technological Situation

Contemporary technology and the present direction of research and development are characterized by intense emphasis on speed and miniaturization, and by the almost complete dominance of digital computing technology. This technology is not just portable, but also mobile, going both inside and outside, and traveling with and inside of the human body both as enhancement and as appendage. Gidden's "time-space convergence" (1995, p. 40) and Harvey's "Time/space compression" (1999) concepts evolve out of the unprecedented emphasis on speed of both information processing and transportation. The result is not simply the reestablishment of an equally relational time/space such as was confounded by the telegraph and "time/space distanciation" (Carey, 1989; Giddens, 1995), but rather, as some argue, that time and space in the contemporary situation are being obliterated entirely as useful concepts (Bauman, 2000; Castells, 2001; Virilio, 2000). The speed and miniaturization of technology and, ultimately, the intense mobility this affords are thus identified as ontological and everyday conceptual crises by many. Giddens (1995) uses a notion of "ontological security" to describe the subjective experience of intensified mobility. The crisis induced by contemporary technology is considered a damaging disruption in the way one is able to know oneself and feel secure in the world. Although this contemporary situation causes concern for some, it has not come about without its historically located prophets.

Although the present circumstances may indeed take time-space compression to unprecedented levels, McLuhan envisioned that this was to come quite clearly as long ago as 1962 (1995). Equally, it is well known that the severe miniaturization and mobility of technology was prophesized quite clearly by Vannever Bush as early as 1945. Although he erred in predicting the dominant medium—giving preference to his own analogical technological development program over the infant digital computer that had been so recently conceived in his time—his vivid description of the future of technology paints a portrait remarkably similar to self-proclaimed cyborg Steve Mann.

Mann himself represents a technologically mobile subject perhaps decades ahead of the masses by dressing himself in his technological harness to mediate his everyday life since the late 1970s. Advances in Bluetooth technology take Mann's experience to an even further frontier, unwiring even the components of one's technologically mobile apparatus: headphones are unwired from their iPods, and control interfaces unwired from the devices they operate, perhaps to be operated even by the mere blinking of one's eyes.

Adding to the confusion of the ontological adjustment accompanying the time/space compression is the potential for ubiquity of human "selves" in time and space. In an allusion to a McLuhanesque "global villiage," contemporary technology creates the circumstances for what Virilio describes as "Tele-presence" (2000). With the compression of history and future into the conceptual present moment, and the obliteration of conceptual spatial limits, the signifiers of selfhood can equally be described as existing everywhere always and nowhere never. Both temporal and spatial distance are confounded as useful ways of contextualizing oneself and the world one lives in. Heidegger's concern that modern technology is dangerous because it "de-distances" is as relevant today as it ever was.

As I've already suggested, although it may well be that contemporary technologies are indeed implicated in an ontological crisis and/or crisis in subjectivity, it is by no means certain that the result has to be entirely negative. Rather than seeing such issues as catastrophic, in a Heideggerian mind-set one might just as well see such ruptures as opportunities to reveal truth, opportunities to glance the essence of "being" itself. Heidegger's concern with "de-distancing" is somewhat distinct from the issues being explored by the authors mentioned above. He is not necessarily concerned with whether or not people can feel comfortable or secure; he is interested rather in ensuring that it is possible for the truth to be revealed. It is the disruption to gathering the elements necessary to create the circumstances in which truth might be experienced that is of main concern as regards technology. As explained in more detail further on, such gathering is possible only through the ability to identify nearness and distance.

Time structures the everyday, and thus a devotion to repairing modernist notions of time towards the goal of ontological security is yet another dedication to maintaining the status quo. Even so, according to Heidegger, we need structures of time and a sense of distance to avoid solipsism and to understand ourselves as belonging in the world. This paradox is inherent in the truth of technology and in part characterizes the essence of being itself. What then of the grander notion I promised to examine in the beginning of this chapter: What is freedom in relation to technology? A thorough review of Heidegger's work provides some insight.

Revealing Heidegger

Heidegger takes up the issue of freedom early in the introduction of his essay "The Question Concerning Technology" (1977). He describes that freedom in relation to technology means understanding its essence as part of our human existence. That is, to recognize both ourselves as technological creatures, and technology as profoundly human. Although he affirms that technology is fundamentally instrumental, "a means to an end" (p.5), that in itself doesn't describe the essence of technology. Rather, the essence of technology can be known only by the work it does, which is to participate in the circumstances of causality.

Heidegger identifies four modes of setting causality into motion, or "occasioning" (p.10), which he describes as entailing (1) the material of which something is produced, (2) the form (or shape) of the phenomenon, (3) the final end use (determining its design), and finally (4) the crafting of the thing (the work that brings it about). These four modes of causality describe parts of the phenomenon of "poiesis," which is "bringing forth" (p.10). Bringing forth is, in its turn, every occasion that something "presences" (makes itself or is made present), which is grounded in "revealing" (p.12). Here, we finally arrive at the work of technology: Presencing.

Technology as a "mode of revealing" (p.13) presences, presencing being truth momentarily illuminating to be seen by humans. To see truth is to see not just ambiguity but also the danger of having that ambiguity kept hidden. In simpler terms, technology is an instrument of causality, and its work is to cause truth to be revealed. Even so, the work of technology is not the essence of technology. Rather, technological essence is enframing, which belongs within the rubric of destining. Although it is the essence of technology, enframing is not technological. However, by revealing humanity as set upon (set in order as part of the standing reserve), enframing is the challenge that motivates technological activity; it is "the real" revealing itself as "standing reserve" (p.23), a conceptual realm within which humans might relegate themselves along with the rest of the natural universe. The nature of this instrumental essence presents the real danger of technology for human "being"—the blind participation in an unending project of ordering the natural world as standing reserve that does not reveal the essence of being but rather keeps one from being able to see it at all:

> Since destining at any given time starts man on a way of revealing, man, thus under way, is continually approaching the brink of the possibility of pursuing and pushing forward nothing but what is revealed in ordering. (Heidegger, 1977, p. 26)

Distinct from premodern technology, modern technology/revealing/enframing is intensely concerned with "setting-in-order," putting the conceived stockpile of nature "on call" (pp.15–17). This standing reserve is a purely conceptual externalization (or perhaps rather a bringing into) of "things" within a closed system of ordering (i.e., an enframing). As an appendage of enframing, theory sets up nature as the standing reserve. This endless cycle of reproducing the same type of enframing as a means of creating a stockpile of nature treats God as a foundational causality (the "first" cause that is otherwise unexplainable) and traps humanity itself in the standing reserve. Thus trapped and unable to see anything outside oneself, humans believe themselves to be the *only* source of agency, triggering a solipsistic slide into the lofty delusion that humans have created and maintain the entire universe. This itself hides the fact that people thus trapped are blinded to their encountering of the actual essence of Being everywhere and anywhere:

> Man stands so decisively in attendance on the challenging-forth of Enframing that he does not apprehend Enframing as a claim, that he fails to see himself as the one spoken to, and hence also fails in every way to hear in what respect he ek-sists, from out of his essence, in the realm of an exhortation or address, and thus *can never* encounter only himself. (Heidegger, 1977, p. 27)

The universe is active, eternally calling out, but falling upon human ears that cannot understand its language, and thereby trapping humanity in its own isolation. For Heidegger, human thinking is reactionary, responsive to the presencing of something. People are "ordered" by enframing to act, primarily acting to drive technology forward. Elaborating the famous "always already" descriptor one often finds references to in feminist and race theory, history is written as destiny, an "unalterable course" (p.25) objectifying time in such a way as to make it accessible to be worked upon. Technology is fateful in this way—the unalterable course of humanity. Humanity is thus the engine of technology, the energy that drives technology forward, participating in poeisis by enframing (rather than understanding the truth of) the universe that presents itself to us. However, it is because of enframing that there exists the potential to learn the truth. The contradiction of enframing what presences itself ("unconceals" itself) is the point at which the opportunity to find truth is found.

In Heidegger's paradigm, human freedom is not about will or causation. Rather, freedom is about being in the realm of destining in such a way that one is able to experience without being "constrained to obey" (p.25), such obedience limiting conceptual possibilities by allowing reproduction of only what is already caused to be. Such constraint is the trap that makes technology dangerous. Freedom is thus not about being free of technology, but rather about the concealing of truth in such a way that the truth shines through and lets the enframing be seen as enframing—metaphorically, the creation of a curtain with peep holes.

Although enframing is dangerous, and, as the essence of technology, it is our fate, humanity is *not* "helplessly delivered over to technology" (p.37). Rather, the way Heidegger describes it, *only* understanding the real nature of our relationship with technology can save us from resigning ourselves to being merely part of the standing reserve. We can be saved by our own active role in technological apparatuses, being agents in our own history as we write it by looking past the surface of technology, refusing to halt investigation at the level of merely the most obvious instrumentality, by going beyond descriptions of only the evident gadgetry, and by understanding technology's essence is enframing. Such an approach demands that we think differently and pay attention to presencing and revealing. Seeing the paradox of the essence of technology as both limiting and libratory allows one to glimpse truth (i.e., the danger of a self-reproductive, self-legitimating enframing, which is in its essence something entirely different).

Truth is revealed through poetics, dwelling in the realm of art. Heidegger describes a poetic "special moment" in destining when the thing that is destined is sent into another destining (p.37). The essence of destining is the destining of Being, while "Being" is the essence of technology (p.38), another aspect that fuels the paradox of technology whose essence is both enframing and Being, since these two essences do not necessarily entail one another, though they are both necessary essences for technology. Humanity is *not* the master of Being, it rather participates in producing and maintaining Being. Being itself is knowable in that special moment of destining, the moment of passage from one destining to another. For Heidegger, this is the experience that the poetics of art is meant to bring about. The essence of technology is the revelation that ordering (structure/pattern, and, paradoxically, clarity) is a frame that hides truth (essence), which is by nature paradoxical and ambiguous, and art is the revelation of the struggle between emptiness and fullness, an experience of the ambiguity of signification.

Truth is revealed only through technological construction. There can be no truth perceivable to humans without technology, where technology is understood as Being, since the concept itself is ideal and needs demonstration through revelation. Techniques of self-discovery, methods of intellectual inquiry, ritualistic and ceremonial behaviors are all understood as part of bringing forth...perhaps the ultimate artwork, the presencing, bringing forth, revelation, and poeisis of oneself. In this process, "Man is...needed and used for the restorative surmounting of the essence of technology" (p.39). To do this, people must first be in a spatial position to carry out this task by attending to their relationship to "Being" (p.40). Attendance to the ontological question requires that one's way of thinking—the way language is used to enframe Being—be such that a solution is actually possible. It is this ability to think differently, to use language differently, to enframe the issue in a way that understanding is achievable, that is at risk.

The danger of enframing is the turning in of thought onto itself, solipsistically enchanted by itself, thereby concealing the truth of Being. In this scenario, enframing is mistaken for Being, entrapping the truth of Being in oblivion. In oblivion, Being is not itself in danger, rather it is kept safe while the world unfolds as it happens without one's active human awareness of Being, or without the desire to know the truth of being that seems to abide outside oneself, instead of the intimate experience of tending the truth of Being. Thus the job of an enlightened humanity is to attend to the truth of Being: "Only when man, as the shepherd of Being, attends upon the truth of Being can he expect an arrival of a destining of Being and not sink to the level of a mere wanting to know" (p.42). The goal is not to bring about Being (which already is), but rather to bring about understanding of Being's truth: to successfully challenge one's enframing—one's mode of thinking—in such a way that the truth about Being is understood. Being itself is not worked upon, not dangerous nor in danger, not touchable except through witnessing the essence of itself: "Being...is not brought about by anything else nor does it itself bring anything about" (p.44).

Technology—enframing—is the same as the coming to presence of

Being. What "is" is "what expressly dwells and endures as present in the 'is' " (p.44). What is presences as what "is," which is the revealed essence of Being, a flash of the truth of Being, like glancing a single spark in a raging fire. That which is is not the same as that which "is," since is remains part of the disguise of enframing, the disguise of the disguise. Revealing what is reveals truth: "Insight into that which is—thus do we name the sudden flash of the truth of Being into truthless Being" (p.47). This is the revelation of humanity itself, an insight in which one is able to behold the nature of the essence of oneself. Such realization has the potential to turn one from looking inwardly and solipsistically, to inspire renunciation of the self-will, and turn oneself towards selflessness and, ultimately, outwards towards immortal divinity.

Thus it is the unexamined devotion to ordering nature into standing reserve—the "technological behavior" of merely enumerating the infinite possible arrays of theorizing nature—that results in "injurious neglect," since simply organizing the world as such results in writing and maintaining only a foundationless history without truth. It is *non*technological behavior—a non-obeying poetic dwelling that inspires a different poiesis, paying attention to presencing and revealing—that has the potential to enframe in such a way that the concept of standing reserve is not merely disrupted but has its essence revealed in a Heideggerian sense: that is, to reveal the moment of passage from one revealing to another; metaphorically, to stand for a moment without citizenship in a border area between nations, unfettered by mortal human dwelling in such a way that one can momentarily see the flash of the immortal as if suddenly glimpsing the infinite cosmos from this ambiguous vantage point, making one aware of the pettiness of this mortal struggle between nations, having glimpsed the divine truth of Being from its safe obscurity in oblivion.

Heidegger's description of our relationship with technology goes a long way to helping understand what is entailed in overcoming the circumstances that would keep humanity obedient to the perpetual motion of Enframing as it reproduces itself, recreating our technologies without hope for freedom. However, the nature of the contemporary danger brought about with the intensification of mobility remains to be further elaborated, as does the nature of poetic dwelling. However, before going any further into these specifics, it will be useful to better understand Heidegger's concept of "pure being," or da-sein.

The purpose of Heidegger's phenomenology is to demonstrate ontology by bringing together the phenomenon—"the self showing in itself"—with logos (scientific inquiry), "a specific mode of letting something be seen" (Heidegger, 1996, pp.27–28). Demonstrating ontology is a means of revealing pure being ("da-sein"). The investigation of pure being is, in fact, the investigation of one's own being (p.39). Phenomenology assumes that people are not objects, but rather "rational life," existing through the deliberate performance of intentional acts, which are meaningfully connected to one another, creating a "unity of a meaning" (p.45). Ontology is foundational to whatever empirical information is crafted. That is, as touched on above, pure being and one's own being are "always already" preexisting any undertaking or scientific investigation. The point one might take from this is, then, that what can be shown in investigative literature is an aesthetic construct that can lead one to *experience.*

The world is ontic and ontological, a type of pure being in itself to which pure being belongs, but world is not a domain (limited). Rather, spatiality (not limited) grants the world existence. The world surrounds being, but being happens in space, which is in turn comprised of a "unity of places" (p.97). Places are the parts of space that belong to being. It is here that the crisis induced by modern technology begins to be better understood.

The speed of modern technology leads to de-distancing of the world, both expanding and destroying the everyday world that surrounds. De-distancing is not only a spatial issue. Being is temporally grounded, in that time structures the world, ordering the everyday. The everyday is what is familiar, by which we interpret and estimate our worldly experiences, and it is this everyday experience that is problematized by the de-distancing of the world. The everyday is reordered in such a way that everything becomes distanceless. This reordering potentially has gravely solipsistic results, in that things can no longer be conceived of as distant from oneself.

In terms of social connectivity, Heidegger asserts that "One belongs to the others oneself" (1996, p. 119). The others—"They"—are the indeterminate, fluid mass of others, represented as a dictatorship of experience—the commonality of experience that produces the "averageness" of the everyday. Like the world, "They" is existential. It is because "they" exists that authenticity is possible. The authentic self is "the self which has explicitly grasped itself" (p.121), which is not the self of the everyday, but which is possible to know by distinction from one's everyday self. Subjectivity exists dispersed in the everyday "they," as a recognition of oneself as "nobody" in the sense of being part of the faceless multitude of "they." Thus it is the modern project of ceaselessly enframing a de-distanced world that endangers the potential for knowing ones own authenticity. Having dissected the problem technology presents us with, let me now turn to Heidegger's proposed solution.

Poetry and Art: Disobeyant Moments

Poetry that thinks is in truth the topology of Being.
This topology tells Being the whereabouts of its actual presence.
—MARTIN HEIDEGGER

Art is a perpetual cycle of a closed paradox. Creation is the causing of things to emerge, where Art is a specific type of creation in which comes to be the thing plus what it reveals: "the becoming and happening of truth" (Heidegger, 2001, p. 69). For Heidegger, art is the truth of being setting itself to work, but in this is a paradox. Art is an origin in its nature. The origin of the work of art is the interrelation of the artist with the work, paradoxically preceded by art itself: "The artist is the origin of the work. The work is the origin of the artist" (p.17). This paradox is perpetually cyclical, and consequently art is the ambiguity of this paradox, to which is added the struggle of the world and the earth, form and material. This truth, which is also untruth, happens in work.

Beyond the "thing" around which assemble what are needed to actualize the product of art, the work art does upon its creation reveals something beyond the obvious form and matter. Abstraction is necessary in order to allow a thing to exist in its own conceptual context/situation/form. Such isolation is possible only in abstraction, since form and matter comprise a "conceptual schema," not able to withstand one another's aesthetic and representational contradictions: "Form as shape is not the consequence of a prior distribution of the matter. The form, on the contrary, determines the arrangement of the matter" (p.27). Form also determines the selection of the type of matter. Shape is, in fact, a consequence of "usefulness" (p.28), towards the fulfillment of a specified purpose. Matter conceptually dissolves into form and the usefulness of equipment. For example, metal may become no longer metal, but rather steel, or an axe.

Things that are created include, in the highest order, "equipment" (p.28), which is useful and artful, but not fully art. Equipment lacks the self-sufficiency of art, the ability to stand and work on its own to reveal something other than its own obviousness. Equipment is made by humans and acts as intermediary between beings and work. The example Heidegger uses is shoes, which, in the context of everyday life, are not art, but which can become equipment used to produce art. The making of equipment is differentiated as craft, not art, which reveals the truth about the equipment: "Art is truth setting itself to work" (pp.35–36). Restated, the "work" of art is the reproduction of a thing's essence, which art works on its own to accomplish, the artist being of no consequence once the art is imbued with the ability to work on its own.

Art is a deployment of work in the present, confronting the past with the present, as in linguistic transformation in definition, setting up a world ("worlding"), conceptually spatializing and thereby liberating "the Open" (p.44). In other words, ordering the world as potentiality rather than as an enclosed domain, and thereby rescuing the open from oblivion, making it known as open. Setting up a world sets forth the earth from the world, parameters establishing other parameters in such a way as it is possible to conceive of there being a world, and that world stands in relation to the earth itself. World confronts earth and vice versa, form versus matter, matter versus form, striving against one another in an effort to assert their natures. Work consists of this struggle of assertions between world and earth. Work that stands on its own exhibits this struggle between strivings, allows one to consider the entirety rather than to glimpse only its parts, and thereby reveals truth.

But this clarity can also hide the essence of being, by lack of knowledge to understand, or deception. This is a reiteration of the paradox of truth. This doesn't negate truth but rather describes how truth always entails its opposite. The struggle to unconceal always entails concealment: Openness happens amidst closures. Beauty can exhibit truth as unconcealed. Truth, therefore, happens in the work of "the work," the nature of work being the struggle as explained. The embedded agitation of the struggle between world and earth is what distinguishes a work as independent, as something that stands on its own. Embedding this agitation is an act of creation, bringing forth the truth.

As mentioned earlier in this chapter, Heidegger identifies at least five ways that truth becomes, all of which would fall under the rubric of art in that the struggle between world and earth is agitated: (1) "truth setting itself into work," what I think of as creative poetic art, or art as artifact; (2) the political act of founding a state, or what I suggest is creation of institutionalizing social systems; (3) "being that is most of all," what I interpret as numinous being, or perhaps more clearly, spirituality; (4) sacrifice, which is a giving over, or surrender to the struggle; (5) a priori thought, which names Being in its questioning of being (2001, p. 60). Although each is distinctly differentiated, these five artful activities share the moniker of "Createdness" (p.60), imbuing them with the strength of the work's ability to work on its own without the artist. The truth in the work—conflict—produces a rift that acts as the intimate bond between world and earth, enacting the paradox of revealing and concealing. This rift is set back—and thereby made observable—by figure/shape. The rift is structured through the figural, which joins the elements necessary for the truth to manifest. Equipment is employed to structure and shape the work, but once set in motion, the struggle of the art allows it to work on its own.

Although artistic work has self-sufficiency in its internal struggle, it needs to be not only created but also preserved. Preservation means knowing by one's being within the struggle of the work. Heidegger describes this as an ontological compulsion, a willingness to transcend, to willingly strive to know the essence of being, and thereby preserve by "standing-within" the struggle of the art (p.65), what I described earlier as standing between borders. Establishing and preserving truth (entailed in the creation of the thing itself) are the two crucial nonmaterial elements that comprise art: "creative preservation" (p.69). As the engines of technology, humans both create and preserve, to reveal the truth, which is the essence of being, and to know that being is the truth not just of one's own being, but of all being.

Poetry is the "saying of world and earth, the saying of the arena of their conflict" (p.71), and poets are the sayers, saying the truth in such a way that their song is existence itself, their own selfness sublimated to the expression of the Being of all things. The foundation of truth, poetry is the essence of art. Modern subjectivity mistakes creation for genius, when the true genius does not create but rather reveals what already is by participating in poiesis. To revisit an earlier statement, the genius of the poets is in their ability to be in the realm of destining in such a way that they are able to experience *without being constrained to obey*. Art is historical, the history of people being the history of creation and preservation, birth and sustenance. History is the experience of being, but not being itself: "Everything is an experience" (2001, p. 77), but such experience is mortal, and it thus expires, unfolding as the slow death of art. Thus, art dies when it ceases to work, when it fails to any longer gather people into its internal struggle, dying with the expiration of witnesses to the struggle within, instead quietly allowing the conflict between world and earth to be gathered somewhere else.

Here we return to the contemporary technological dilemma. Although technologies end remoteness, they do not actually make things nearer or closer. The original distance is embedded in the media themselves. So the problem intensified with de-distancing technologies, then, is that it keeps things from being brought near. For example, television brings remote things within view but at the same time holds them away and keeps them from actually being closer. Instead of bringing far things near, everything becomes part of a "uniform distancelessness" (p.164). Nearness itself is the result of attending to what is near—one can be near their television, but not generally near to what is shown by the television as distant. Nearing is the presencing of nearness, which preserves farness. A thing, be it equipment or art, is something that does what work it is meant to do, thereby presencing by gathering what it gathers to do its work, making it what it is, which is a thing. Heidegger describes earth, sky, divinities, and mortals as the four interrelated aspects of the world worlding, reflecting themselves in the perceived unity of the world gathering the elements that it needs to be the world. His taxonomy aside, gathering and thereby bringing elements together is necessary for things to presence. Hence the abolition of distance (nearness and farness) does not allow elements to be gathered for the presencing of things in the sense of their eruption out of the world, but rather without recognition it relegates the unknown to perpetual safekeeping in oblivion.

Towards a Cultural Approach to Mobile Technology

We live in a universe that is animated and has agency of its own in spite of human efforts to suspend its animation by ordering it into the standing reserve: "The constellation of Being is uttering itself to us" (Heidegger, 1977, p. 48). We can establish the circumstances in which to hear this uttering by daring to stand inside the conflict that is the work of truth—to stand within the rift that bonds world and earth, and through poiesis, participate in enacting the paradox of revealing and concealing. The enactment of this rift necessitates incorporating the contributions of "actant" technologies that, as equipment, also participate in the gathering. Only by recognizing all the elements implicated in the struggle between earth and world can we have a free relationship with technology.

Human freedom is commonly romanticized in Western thought as a self-decided/self-directed course of action in individuals' lives, yet by following Heidegger's reasoning we find that in truth it is in its essence yet another program of action, perhaps not self-decided or self-directed at all. Although undoubtedly more desirable than its negative counterpart, the popular discourse of "freedom" is another construct within which people are situated. By elaborating the necessity of enframing for function in everyday life, but yet defining humans as the driving force of technology, Heidegger eradicates the false dichotomy that spoils the philosophical grounds of so many others regarding human/machine relations. As such, his work suggests a theoretical paradigm that can be worked into a methodology capable of describing medium-specific human/machine phenomena and interactionary experiences—for my purposes, technography.

Heidegger's philosophy of technology is broad enough to incorporate sophisticated nuances and subtleties associated with contemporary inventions and their "creation, apprehension and use" (Carey, 1989), while at the same time allowing for a great amount of specificity in any direction one chooses to emphasize. Pacey (1996) includes "goals, values and ethical codes, belief in progress, awareness and creativity" (p.6) as cultural aspects of technology, all of which can be known through the paradigm of qualitatively examining and recreating the experiences of technologically induced disjunctures. A method rooted in Heidegger's philosophy of technology should not only investigate the artful but also be artful in its reporting. That is, it should not merely report but illuminate by its own example, to gather the elements of the world/earth conflict together and allow individuals to thus stand inside of the truth. I suggest that this should be the foundational criteria for a method of technography from a qualitative research approach, incorporating elements from the contemporary frontiers of ethnography that recognize that we are recreating culture even as we write it (Denzin, 2003). However, to cause the accurate reproduction of authentic technological experiences and thereby reveal the truth about them, techniques borrowed from autoethnography and performance ethnography must be imbued with the struggle that is appendicized into the programs of our technologies and must contend with the distance embedded in the media that de-distances the world (see Kien, 2005).

If to be human is—in the way I have agreed with Heidegger—to be profoundly miscegenated with technology, if human freedom is relational to one's agency in nudging the direction of technological production and reproduction towards the revelation of pure being, there can be no better way forward than embracing the disobedient idea that inducing some discomfort and struggle with not just technology, but all practices of everyday life, might be a good thing that allows for an opening of the normative enframing of the everyday. This can allow the turning of one's gaze outwards, towards something greater than the solipsistic habit in which many are now engaged. Most importantly for my own ethical imperative, this is an approach that can identify practices of freedom from potentially anyone, anywhere, not limited only to hi-tech nomads. The reality of everyday life is, after all, such that it is often much easier and cheaper for non-elites to acquire guns than computers, but the acquisition of more guns obviously doesn't guarantee any further freedom for humanity. Rather, it is only by being able to stand inside such conflict without being conceptually constrained to obey that our shackles (metaphorical and otherwise) can be revealed as mere shackles.

Glossary of Actor-Network Theory Terms

Actant Performing inscribed roles, actants are the primary unit in building networks. Actants assemble relationships by entering into alliances that in turn form network arrays. An actant may appear stable or dominant due to the inability of other actants within its network to build oppositions (Latour, 1988; Callon and Law, 1995).

Actant-Network A spatializing array of actants performing their inscribed roles to create and participate in alliances.

Actor A type of actant that has network distributions thrust upon it, with which it struggles for definition. An actor is the temporal sum total of all the relations that compose it (Law, 1992; Callon and Law, 1995).

Actor/Agent/ Actant Having the key feature of autonomy, these terms are interchangeable and may be used to refer to any individual, collective, figurative, or nonfigurative entity (Latour, 1988b).

Actor-Network A spatializing array of actants that includes actors in contestation to turn the network(s) they are constituted by towards the fulfillment of their own motivations and translations.

Actor-Network Theory (ANT) The connection of researchers with actors, with the intention of summing up interactions through various kinds of devices, inscriptions, forms, and formulae, into very local, very practical, very tiny loci. ANT seeks to describe what provides actants with their actions, subjectivity, intentionality, and their morality, recording how worlds are built in specific sites (Latour, 1999).

Agent An actant located in an area that has agency allocated to it by a Collectif (Callon and Law, 1995).

Agnostic Observation A research perspective characterized by the avoidance of judgments and interpretations of observed phenomena, believing that foundations and absolutes are not possible (Callon, 1986).

Alliance The enlistment of one or more actant by another, which is the basis of network formation and maintenance (Latour, 1988b).

Ally An actant that has been enlisted in the constitution of a network. Allies can be coerced, persuaded, willing or unwilling, and even unknowing participants in a network (Latour, 1988b).

Authority Figure An actant that resists individuation in its performance as part of an institution's administrative machinery (Latour, 1988a).

Autonomy Maintenance of the appearance of stability of an actant through the constant building of relationships with other actants (Latour, 1988b).

Collectif, Hybrid An array of relations, links interpenetrations, and processes. A Hybrid Collectif can be contrasted with "a collective," which is a thing. Not all Collectifs can be agents, but some are. A Collectif includes all that inspires, influences, and touches it. In this understanding of agency, differences and dualism are generated out of partial similarities. Collectifs allocate agency to a particular area of their network, and the actants located there are then said to be agents (Callon and Law, 1995).

Competition All actants in an alliance, even when waiting in silence, are motivated to continually seek to divert the network of alliances towards their own ends (Latour, 1988b).

Des-inscription When an inscribed reader breaks from prescribed behavior (Latour, 1988a).

Dominance The appearance of singularity from an actant with the agency to translate and speak on behalf of the entire actor-network. A dominating voice justifies itself as democratic, merely enunciating what the network demands of it, which is in effect quite true, since the dominating voice will say only that which creates new alliances and maintains the network, or it will soon lose its stability and be usurped by another voice (Latour, 1988b).

Enlistment When one actant persuades another into an allied network (Latour, 1988b).

Entelechy A matrix of networked actors and actants that appears as a singular entity.

Fact See Truth.

Free Association Rejection of a priori categorization to allow actors to define and associate the elements of their world according to their own language (Callon, 1986).

Generalized Symmetry Explaining conflicting viewpoints and arguments using the same terms or voice for all points of view, not changing voices to highlight or privilege specific understandings of a phenomenon (Callon, 1986).

Identification The laborious construction and vigorous maintenance of unequal translative exchanges. Identification always demonstrates inequality, with clear winners and losers (Latour, 1988b).

Immutable Mobile A token that circulates as a symbol of power with the ability to travel great distances and endure numerous translations and yet retain much of its original form in spite of being passed from actor to actor (Law, 1986).

Inscribed Reader A person in whom a translation has been inscribed, thus an inscribed reader is prepared to perform the relations the translation has prescribed upon encountering the context described by the translation (Latour, 1988a).

Inscription or Encoding Translation of any script from one repertoire to a more durable one (e.g., spoken words into printed text) (Latour, 1988a).

Lieutenants Machines and machinations that hold the places and perform the roles delegated to them in lieu of a human actor, performing the social relations they are inscribed with (Latour, 1988a).

Locality Centers of translation (Law, 1992).

Location The process of actants finding and being found by one another, in which time and space act as descriptive frameworks but are also hegemonic impositions (Latour, 1988b).

Machination The enactment of a programmatic inscription by an authority figure, as part of institutional administrative machinery (Latour, 1988c).

Machine As actant, part of the political ordering of the world. Machines are inscribed with social codes and enrolled as sophisticated allies in the recruitment of other allies and keeping them in place (e.g., military equipment) (Latour, 1988c).

Messy Network A productive assemblage of individual actors, technology, society, and economics (Bijker and Law, 1992).

Method, Actor-Network The textual provision of time and space to actively negotiate and/or impose definitions and measures, creating the appearance of translation, explanation, understanding, control, purchase, decision, and conviction, with the purpose of enrolling other actors and making them work. In ANT, methodology is understood as an inexact hegemonic process of narrative construction that every other actant understands and thereby gains the appearance as the most powerful translation (Latour, 1988b).

Network A term used to describe series of transformations, translations, and transductions in constant mutation (Latour, 1999).

Network Consolidation The appearance of singularity of a network, since networks are ordered with materials and strategies, creating aesthetic patterns of generated effects of power and hierarchy in institutions and organizations (Law, 1992).

Opposition A rearranged array of alliances that struggles to usurp the array from which it comes (Latour, 1988b).

Performativity The enactment of inscribed codes that makes entities durable and fixed within networks (Law, 1999).

Person An effect produced by a network of heterogeneous, interacting materials (Law, 1992).

Potency An aesthetic spatialization of setting forces against each other, collecting the faithful as an inside pushing against the doubtful (outside) as in the reification of modernist essentialisms (Latour, 1988b).

Power The effect and consequence of performances in networks, produced by associating entities together into networks formed through alliances (as in translation). Power appears to be possessed by individuals through the performance of acts by others on what appears to be the behalf of the dominating actant, or the potential of having others act for oneself. Tokens circulate as symbols of power, appearing to be in the possession of the holder, when, in fact, they must remain in circulation in order to remain potent (Latour, 1986; Law 1992).

Pre-inscription Everything that prepares the scene (the context of performance of relations) for articulation (Latour, 1988a).

Prescription Whatever is presupposed from transcribed actors and authors, including presuppositions encoded in machines (Latour, 1988a).

Principle of Relativity The assertion that relations between actants are always unequal, that everything is relative and imbued with inequality and clear winners and losers (Latour, 1988b).

Reality — Revealed in the practice of relations, reality is demonstrated and measured by lines of force drawn in the performance of relations (Latour, 1988b).

Relational Materiality — Understanding form as an effect of relations, made durable and fixed through performance (Law, 1999).

Reliability — The extent to which an alliance is reliable is the extent to which a critical mass of other actants in the alliance say for themselves they want the same thing as the dominating voice (Latour, 1988b).

Repertoire — Translation of a performative script from a human body to a machine, or vice versa, entailing inscription/encoding (Latour, 1988a).

Singularity — The appearance of network consolidation, as if the network were a single entity (i.e., a single actant) in and of itself, working towards a single goal or on behalf of a singular interest (Law, 1992).

Social — The patterned networks of heterogeneous materials composed of people, machines, animals, texts, architecture, and symbolic material (such as money, language, etc.) (Law, 1992).

Society — The performance of relationships. A definitive list of the properties of society is theoretically impossible, but it is possible in practice by listing what is enacted in everyday performance (Latour, 1986).

Sociologism — The ability of one to read the scripts of nonhuman actors (Latour, 1988a).

Space Actor-Networks are geometric spatial arrays, non-Euclidian in nature (i.e., they do not exist as a reality "outside" oneself). Like time, space is a hegemonic imposition in the form of a descriptive interpretive framework (Latour, 1988b).

Spatialization The imposition of a hegemonic descriptive interpretive framework.

Strategy The maintenance of network stability through time, strengthened through the embodiment of relations in durable materials (Law, 1992).

Subscription When an inscribed reader acquiesces to prescribed behavior (Latour, 1988a).

Technologism The ability of humans to read their own behavioral scripts as required by the technology they interact with (Latour, 1988a).

Time Actants struggle for temporal dominance. Like space, time is a hegemonic imposition in the form of a descriptive interpretive framework (Latour, 1988b).

Token An object or objectified conceptual symbol fetishized and reified as a container of power. Tokens are set into motion and travel through networks of actors that propel them until they are met with resistance, as in Newton's law of inertia. In the process, actors may reshape and transform the token (translate it) as they pass it along (Latour, 1986).

Translation The transformation of any script from one repertoire to a more durable one. Translation produces the appearance of power and knowledge through the aesthetic expression of a process in which one actor assumes to speak on behalf of all the other actors that constitute an alliance. In practice, translation projects a deceptive appearance of essential definitions, creating circumstances that give rise to illusions of power and control (Callon and Latour, 1981; Callon, 1986; Latour, 1988a).

Truth A declaration whose alliances are intact. A sentence with insufficient allies appears false (Latour, 1988b).

Bibliography

Althusser, L. (1971). "Ideology and Ideological State Apparatuses (Notes towards an Investigation)." *Lenin & Philosophy and Other Essays.* New York: Monthly Review Press.

Angrosio, M. V. (2005). "Recontextualizing Observation: Ethnography, Pedagogy, and the Prospects for a Political Agenda." *The Sage Handbook of Qualitative Research.* N. K. Denzin and Y. S. Lincoln (eds.). Thousand Oaks, CA: Sage, pp. 729-746.

Appadurai, Arjun. (2001). "Grassroots Globalization and the Research Imagination." *Globalization.* Arjun Appadurai (ed.). Durham, NC: Duke University Press, pp.1–21.

Arnold, M. (2003). "On the Phenomenology of Technology: The 'Janus-Faces' of Mobile Phones." *Information and Organization,* 13, 231–256.

Barney, D. (2000). *Prometheus Wired: The Hope for Democracy in the Age of Network Technology.* Toronto: UBC Press.

Barry, A. (2001). *Political Machines: Governing a Technological Society.* New York: Athlone Press.

Baudrillard, Jean. (1988). *The Ecstacy of Communication.* Bernard and Caroline Schutze (trans.). Sylvére Lotringer (ed.). New York: Autonomedia.

———. (2002). *The Spirit of Terrorism.* New York: Verso.

Bauman, Z. (2000a). *Liquid Modernity.* Cambridge, MA: Polity Press.

———. (2000b). "Note on Society/Note de société: Ethics of Individuals." *Canadian Journal of Sociology,* 25:1, 83–96.

Benjamin, Walter. (1978). *Illuminations.* Harry Zohn (trans.). New York: Schocken Books.

Bhabha, Homi. (1990). "Introduction: Narrating the Nation." *Nation and Narration.* Homi Bhabha (ed). New York: Routledge, Chapman and Hall, pp.1-7.

Bijker, Wiebe E., and John Law. (1992). "General Introduction." *Shaping Technology/Building Society: Studies in Sociotechnical Change.* Wiebe E. Bijker and John Law (eds.). Cambridge, MA: MIT Press.

Blackhurst, A. E., and Edyburn, D. L. (2000). "A Brief History of Special Education Technology." *Special Education Technology Practice,* 2:1, 21–35.

———. (2001). "Types of Technology." *National Assistive Technology Research Institute.* Online: <http://tam.uky.edu/basics/techtypes.html>.

Blunt, Alison. (2007). "Cultural Geographies of Migration: Mobility, Transnationality and Diaspora." *Progress in Human Geography,* 31:5, 684–694.

Boschken, Herman L. (2008). "A Multiple-Perspectives Construct of the American Global City." *Urban Studies*, 45:1, 3–28.

Burgelman, Jean Claude. (2000). "Traveling with Communication Technologies in Space, Time, and Everyday Life: An Exploration of Their Impact." *First Monday*, 5:3. Online <http://firstmonday.org/issues/issue5_3/burgelman/index.html>.

Callon, Michel. (1986). "Some Elements of a Sociology of Translation: Domestication of the Scallops and the Fishermen of St. Brieuc Bay." *Power, Action and Belief: A New Sociology of Knowledge? Sociological Review Monograph* 32: 196–233. John Law (ed.). London: Routledge & Kegan Paul.

———. (1999). "Actor-Network Theory—The Market Test." *Actor Network Theory and After.* John Law and John Hassard (eds.). Malden, MA: Blackwell, pp.181–195.

Callon, Michel, and Bruno Latour. (1981). "Unscrewing the Big Leviathan: How Actors Macrostructure Reality and How Sociologists Help Them to Do So." *Advances in Social Theory and Methodology: Toward an Integration of Micro- and Macro-Sociologies.* K. d. Knorr-Cetina and A. V. Cicourel (eds.). Boston: Routledge & Kegan Paul, pp.277–303.

Callon, Michel, and John Law. (1995). "Agency and the Hybrid Collectif." *South Atlantic Quarterly*, 94: 481–507.

Campbell, Howard, and Josiah Heyman. (2007). "Slantwise: Beyond Domination and Resistance on the Border." *Journal of Contemporary Ethnography*, 36:1, 3–30.

Carey, James. (1989). *Communication as Culture: Essays on Media and Society.* New York: Unwin/Hyman.

Castells, M. (2000). *The Rise of the Network Society*, 2nd edition. Malden, MA: Blackwell.

———. (2001). *The Internet Galaxy: Reflections on the Internet, Business, and Society.* New York: Oxford University Press.

Castells, Manuel, and Peter Hall. (1994). *Technopoles of the World.* London: Routledge.

Chan, Stephanie. (2008). "Cross-cultural Civility in Global Civil Society: Transnational Cooperation in Chinese NGOs." *Global Networks*, 8:2, 232–252.

Charland, Maurice. (1986). "Technological Nationalism." *Canadian Journal of Political and Social Theory*, X:1.

Chatterjee, P. (1993). *The Nation and Its Fragments: Colonial and Postcolonial Histories.* Princeton, NJ: Princeton University Press.

Christians, C. G. (1997). "Technology and Triadic Theories of Mediation." *Rethinking Media, Religion and Culture.* Stewart Hoover (ed.). Thousand Oaks, CA: Sage.

Coombe, Rosemary. (1998). *The Cultural Life of Intellectual Properties: Authorship, Appropriation, and the Law.* Durham, NC: Duke University Press.

Couch, Danielle, and Pranee Liamputtong. (2008). "Online Dating and Mating: The Use of the Internet to Meet Sexual Partners." *Qualitative Health Research,* 18:2, 268–279.

Cvetkovich, Ann, and Douglas Kellner. (1997). "Introduction: Thinking Global and Local." *Articulating the Global and the Local: Globalization and Cultural Studies.* Ann Cvetkovich and Douglas Kellner (eds.). Boulder, CO: Westview Press, pp.1–32.

Davies, B., Browne, J., Honan, E., Laws, C., Mueller-Rockstroh, B., and Bendix-Petersen, E. (2004). "The Ambivalent Practices of Reflexivity." *Qualitative Inquiry,* 10:3, 360–389.

Davis, M. (1990). *City of Quartz: Excavating the Future in Los Angeles.* New York: Verso.

de Certeau, Michel. (1984). *The Practice of Everyday Life.* Steven Rendall (trans.). Berkeley: University of California Press.

Dell, Peter, and Marinova, Dora (2002). "Erving Goffman and the Internet, Theory of Science (Teorie Vedy)." *Journal for Theory of Science, Technology and Communication,* 4, 85–98.

Denzin, Norman K. (1999). "Cybertalk and the Method of Instances." *Doing Internet Research: Critical Issues and Methods for Examining the Net.* S. G. Jones (ed.). Thousand Oaks, CA: Sage, pp.107–125.

———. (2001). *Interpretive Interactionism,* 2nd edition. Thousand Oaks, CA: Sage.

———. (2003). *Performance Ethnography: Critical Pedagogy and the Politics of Culture.* Thousand Oaks, CA: Sage.

Denzin, Norman K., and Yvonna S. Lincoln (eds.). (2002). *The Qualitative Inquiry Reader.* Thousand Oaks, CA: Sage.

Dholakia, Nikhilesh, and Detlev Zwick. (2003). "Mobile Technologies and Boundaryless Spaces: Slavish Lifestyles, Seductive Meanderings, or Creative Empowerment?" Paper presented at the Home Oriented Informatics and Telematics conference, University of California, Irvine, April 6–8, 2003.

Dupuis, Ann, and David Thorns. (1998). "Home, Home Ownership and the Search for Ontological Security." *Sociological Review,* 46:1, 24–47.

Durkheim, Émile. (1996). "The Elementary Forms of Religious Life." *Readings in Social Theory.* Francis Farganis (ed.). Toronto: McGraw-Hill.

Edensor, Tim. (2004). "Automobility and National Identity." *Theory, Culture and Society*, 21:4/5, pp.101–120.

Ellis, C., and Bochner, A. P. (2000). "Autoethnography, Personal Narrative, Reflexivity: Researcher as Subject." *Handbook of Qualitative Research*, 2nd edition. N. K. Denzin and Y. S. Lincoln (eds.). Thousand Oaks, CA: Sage, pp.733–768.

Featherstone, M. (1991). *Consumer Culture and Postmodernism*. Newbury Park, CA: Sage.

Feenberg, A. (1991). *Critical Theory of Technology*. New York: Oxford University Press.

Fortunati, Leopoldina. (2002). "Italy: Sterotypes, True and False." *Perpetual Contact: Mobile Communication, Private Talk, Public Performance*. James E. Katz and Mark Aakhus (eds.). New York: Cambridge University Press, pp. 42-62.

Foucault, Michel. (1967). "Of Other Spaces, Heterotopias." Translated by Jay Miskowiec from the original "Des Espace Autres" in Architecture/Mouvement/Continuité, October 1984. Online January 2, 2005: <http://foucault.info/documents/heteroTopia/foucault.heteroTopia.en.html>.

———. (1979). *Discipline and Punish: The Birth of the Prison*. Alan Sheridan (trans.). New York: Vintage Books.

———. (1980). *Power/Knowledge: Selected Interviews & Other Writings 1972–1977*. Colin Gordon (ed.). New York: Pantheon Books.

———. (1984). "Space, Knowledge and Power." *The Foucault Reader*. Paul Rabinow (ed.). New York: Pantheon Books, pp. 239–256.

———. (1990). *The History of Sexuality: An introduction, 1*. Robert Hurley (trans.). New York: Vintage Books.

———. (1999). "Space, Power and Knowledge." *The Cultural Studies Reader*, 2nd edition. Simon During (ed.). London: Routledge, pp. 134–141.

Fry, Tony (ed.). (1993). *R U A TV? Heidegger and the Televisual*. Sydney: Power Publications.

Gannon, S. (2001). "Representing the Collective Girl: A Poetic Approach to a Methodological Dilemma." *Qualitative Inquiry*, 7:6, 787–800.

Gauntlett, D. (2002). *Media, Gender and Identity: An Introduction*. New York: Routledge.

Giddens, A. (1986). *The Constitution of Society: Outline of the Theory of Structuration*. Berkeley: University of California Press.

———. (1990). *The Consequences of Modernity*. Stanford, CA: Stanford University Press.

———. (1991). *Modernity and Self-Identity: Self and Society in the Late Modern Age.* Stanford, CA: Stanford University Press.

———. (1995) *Politics, Sociology and Social Theory: Encounters with Classical and Contemporary Social Thought.* Cambridge, UK: Polity.

Gilroy, P. (2001). *Against Race: Imagining Political Culture beyond the Color Line.* Cambridge, MA: Harvard University Press.

Goffman, E. (1959). *The Presentation of Self in Everyday Life.* New York: Anchor.

———. (1963). *Behavior in Public Places: Notes on the Social Organization of Gatherings.* New York: Free Press.

———. (1971). *Relations in Public: Microstudies of the Public Order.* New York: Harper & Row.

———. (1974). *Frame Analysis: an Essay on the Organization of Experience.* New York: Harper & Row.

Gotved, Stine. (2006). "Time and Space in Cybersocial Reality." *New Media & Society,* 8:3, 467–486.

Gunkel, D. J. (2001). *Hacking Cyberspace.* Boulder, CO: Perseus Books.

Hamelink, C. J. (2000). *The Ethics of Cyberspace.* Thousand Oaks, CA: Sage.

Hansen, Miriam. (1993). "Unstable Mixtures Dilated Spheres: Negt and Kluge's The Public Sphere and Experience, Twenty Years Later," *Public Culture,* 5, 179–212.

Haraway, Donna. (1999). "A Cyborg Manifesto." *The Cultural Studies Reader,* 2nd edition. Simon During (ed.). New York: Routledge, 271–291.

Hardt, Michael, and Antonio Negri. (2000). *Empire.* Cambridge, MA: Harvard University Press.

Harvey, D. (1989). *The Condition of Postmodernity: An Enquiry into the Origins of Cultural Change.* New York: Basil Blackwell.

———. (1999). *Spaces of Capital: Towards a Critical Geography.* New York: Routledge.

Hay, James. (1996). "The Place of the Audience: Beyond Audience Studies." *The Audience and Its Landscape.* James Hay, Lawrence Grossberg and Ellen Wartella (eds.). Boulder, CO: Westview Press, pp.359–78.

———. (2001). "Locating the Televisual." *Television and New Media,* 2:3, 205–234.

Hay, James, and Jeremy Packer. (2004). "Crossing the Media(-n): Auto-mobility, the Transported Self, and Technologies of Freedom." *Media/Space: Scale and Culture in a Media Age.* Nick Couldry and Anna McCarthy (eds.). Routledge, New York, pp. 209–232.

Hayles, K. (1999). *How We Became Posthuman: Virtual Bodies in Cybernetics, Literature, and Informatics.* Chicago: University of Chicago Press.

Heidegger, Martin. (1977). *The Question Concerning Technology and Other Essays.* William Lovett (trans.). New York: Harper and Row.

———. (1996). *Being and Time.* Joan Stambaough (trans.). Albany: State University of New York Press.

———. (2001). *Poetry, Language, Thought.* Albert Hofstadter (trans.). New York: HarperCollins.

Held, David, Anthony McGrew, David Goldblatt, and Johathan Perraton. (1999). *Global Transformations: Politics, Economics and Culture.* Stanford, CA: Stanford University Press.

Hine, C. (2000). *Virtual Ethnography.* London: Sage.

Hine, Julian, Derek Swan, Judith Scott, David Binnie, and John Sharp. (2000). "Using Technology to Overcome the Tyranny of Space: Information Provision and Wayfinding." *Urban Studies,* 37:10, 1757–1770.

Hobbes, T. (1968). *Leviathan.* New York: Penguin.

Humphreys, Lee. (2005). "Cellphones in Public: Social Interactions in a Wireless Era." *New Media & Society,* 7:6, 810–833.

Innis, H. (1986). *Empire and Communication.* Victoria, BC: Press Porcépic Ltd.

———. (1999). *The Bias of Communication.* Toronto: University of Toronto Press.

Jones, S. G. (ed.). (1997). *Virtual Culture: Identity and Communication in Cybersociety.* Thousand Oaks, CA: Sage.

———. (ed.). (1998). *Cybersociety 2.0: Revisiting Computer-Mediated Communication and Community.* Thousand Oaks, CA: Sage.

Katz, James E., and Satomi Sugiyama. (2006). "Mobile Phones as Fashion Statements: Evidence from Student Surveys in the US and Japan." *New Media & Society,* 8:2, 321–337.

Kemmis, S., and McTaggart, R. (2005). "Participatory Action Research: Communicative Action and the Public Sphere." *The Sage Handbook of Qualitative Research,* 3rd edition. N. K. Denzin and Y. S. Lincoln (eds.). Thousand Oaks, CA: Sage, pp.559–604.

Khalideen, Rosetta. (2008). "Voiceless in the Internationalized University Classroom: Diversity and the Dynamics of Difference." *The International Journal of Diversity in Organisations, Communities and Nations,* 7:6, 267–274.

Kien, G. (2002). *The Digital Story: Analyzing Binary Code as a Cultural Text.* Master's Thesis. York University, Toronto.

———. (2003). "Portability, Auto-Mobility and Ontological Security." Paper presentation. *90th Annual Meeting of the National Communication Association: 2004 Convention*, Chicago, November 2004.

———. (2004a). "Culture, State, Globalization: The Articulation of Global Capital." *Cultural Studies ↔ Critical Methodologies*, 4:4, 472–500.

———. (2004b). "Technographic Media Studies: A Way Forward in a Distanceless World." Paper presentation. *New Forms Festival 20IV: Technography*, Vancouver, Canada, October 2004.

———. (2005). "Internet Time: Socio-Spatial Coordination Online." AoIR Researcher Annual, Volume 2. New York: Peter Lang, pp.16–30.

———. (2006). "Postmodern Gargoyles, Simulated Power Aesthetics." *Qualitative Inquiry*, 12:3, 681–703.

———. (2007). "A Western Consumer in South Korea: Autoethnographical Fiction of Western Performance." *Cultural Studies ↔ Critical Methodologies*, 5:3, 264–280.

———. (2008). "Technography = Technology + Ethnography." *Qualitative Inquiry*, 14:7, 1101–1109.

———. (in press a). "Virtual Environment: The Machine Is Our World." *Identity, Learning and Support in Virtual Environment.* Sharon Tettegah and Cynthia Cologne (eds.). Waganingen, The Netherlands: Sense.

———. (in press b). "Hybrid Networks, Relational Materiality." *Material Culture and Technology in Everyday Life.* Phillip Vannini (ed.). New York: Peter Lang.

Kroker, Arthur. (2002). "Hyper-Heidegger." *Ctheory.net.* Arthur and Marilouise Kroker (eds.). Online: <http://ctheory.net/printer.asp?id=348>.

Latour, Bruno. (1986). "The Powers of Association." *Power, Action and Belief: A New Sociology of Knowledge? Sociological Review Monograph* 32: 264–280. John Law (ed.). London: Routledge & Kegan Paul.

———. (1988a). "Mixing Humans and Nonhumans Together: The Sociology of a Door-Closer." *Social Problems*, 35:3, 298–310.

———. (1988b). *The Pasteurization of France.* Alan Sheridan and John Law (trans.). Cambridge, MA: Harvard University Press.

———. (1988c). "*The Prince* for Machines as well as for Machinations." *Technology and Social Process.* B. Elliott (ed.). Edinburgh, UK: Edinburgh University Press, pp.20–43.

———. (1992). "Where Are the Missing Masses? The Sociology of a Few Mundane Artifacts." *Shaping Technology/Building Society: Studies in Sociotechnical Change.* Weibe Bijker and John Law (eds.). Cambridge, MA: MIT Press, pp. 225–258.

———. (1998). "On Actor-Network Theory: A Few Clarifications." Centre for Social Theory and Technology, Keele University. Online December 22, 2004, <http://amsterdam.nettime.org/Lists-Archives/nettime-l-9801/msg00019.html>.

———. (1999). "On Recalling ANT." *Actor Network Theory and After.* John Law and John Hassard (eds.). Malden, MA: Blackwell, pp.15–25.

Latour, Bruno, and S. Woolgar. (1979). *Laboratory Life: The Social Construction of Scientific Facts.* Beverly Hills: Sage.

Law, John. (1986). "On the Methods of Long-Distance Control: Vessels, Navigation and the Portuguese Route to India. Power, Action and Belief: A New Sociology of Knowledge?" *Sociological Review Monograph,* 32. John Law (ed.). London: Routledge & Kegan Paul, pp.234–263.

———. (1992). "Notes on the Theory of the Actor-Network: Ordering, Strategy, and Heterogeneity." *Systems Practice,* 5:4, pp.379–393.

———. (1999). "After ANT: Complexity, Naming and Topology." *Actor Network Theory and After.* John Law and John Hassard (eds.). Malden, MA: Blackwell, pp.1–14.

———. (2000). "Objects, Spaces and Others." Center for Science Studies, Lancaster University. Online December 26, 2004, <http://www.comp.lancs.ac.uk/sociology/papers/law-objects-spaces-others.pdf>.

———. (2002). *Aircraft Stories: Decentering the Object in Technoscience.* Durham, NC: Duke University Press.

Lee, Kwang-Suk. (2008). "Globalization, Electronic Empire and the Virtual Geography of Korea's Information and Telecommunications Infrastructure". *The International Communication Gazette,* 70:1, 3–20.

———. (2005). "Neuroticism: End of a Doctoral Dissertation." *Qualitative Inquiry,* 11:6, 933–938.

Lefebvre, Henri. (1991). *The Production of Space.* Donald Nicholson-Smith (trans.). Oxford, UK: Blackwell.

Licoppe, Christian, and Jean-Philippe Heurtin. (2002). "France: Preserving the Image." *Perpetual Contact: Mobile Communication, Private Talk, Public Performance.* James E. Katz and Mark Aakhus (eds.). New York: Cambridge University Press, pp. 94–109.

Ling, R. (2004). *The Mobile Connection: The Cell Phone's Impact on Society.* San Francisco: Elsevier/Morgan Kaufmann.

Lippman, Walter. (1925). *The Phantom Public.* New York: Harcourt

Locke, John. (1980). *Second Treatise of Government.* C. B. Macpherson (ed.). Indianapolis, IN: Hackett.

Lunenfeld, Peter (ed.). (2001). *The Digital Dialectic: New Essays on New Media.* Cambridge, MA: MIT Press.

Lury, C. (1996). *Consumer Culture.* Cambridge, MA: Polity Press.

Lyotard, Jean-François. (1991). *The Inhuman: Reflections on Time.* Geoffrey Bennington and Rachel Bowlby (trans.). Oxford: Blackwell.

Mann, Steve, and Hal Niedzviecki. (2001). *Cyborg: Digital Destiny and Human Possibility in the Age of the Wearable Computer.* Toronto: Random House.

Markham, A. (1998). *Life Online: Researching Real Experience in Virtual Space.* New York: AltaMira Press.

Massey, Doreen. (1993). "Power-Geometry and a Progressive Sense of Place." *Mapping the Futures.* Jon Bird, Barry Curtis, Tim Putnam, George Robertson, and Lisa Tickner (eds.). London: Routledge, pp. 59–69.

———. (1994). *Space, Place and Gender.* Minneapolis: University of Minnesota Press.

———. (1999). "Imagining Globalisation: Power-Geometries of Time-Space," "Philosophy and Politics of Spatiality: Some Considerations." *Global Futures: Migration, Environment and Globalization.* A. Brah, M. Hickman, and M. Mac an Ghaill (eds.). Basingstoke, UK: Macmillan, pp. 27–44.

Massey, Doreen, and Pat Jess. (1995). *A Place in the World: Places, Cultures and Globalization.* New York: Oxford University Press.

Mattelart, Armand. (2000). *Networking the World, 1794–2000.* Liz Carey-Libbrecht and James A. Cohen (trans.). Minneapolis: University of Minnesota Press.

McCarthy, C. (1998). *The Uses of Culture: Education and the Limits of Ethnic Affiliation.* New York: Routledge.

McCarthy, Cameron, Michael D. Giardina, Susan J. Harewood, and Jin-kyung Park. (2003). "Contesting Culture: Identity and Curriculum Dilemmas in the Age of Globalization, Postcolonialism and Multiplicity." Revised Draft. First published in *Harvard Educational Review,* 73:3, 449–465.

McChesney, Robert, Ellen Meiksina Wood, and John Bellany Foster (eds.). (1998). "The Political Economy of Global Communication." *Capitalism and the Information Age.* New York: Monthly Review Press, pp.1–26.

McLuhan, M. (1964). *Understanding Media.* Mentor/Penguin.

———. (1995). *Essential McLuhan.* Eric McLuhan and Frank Zingrone (eds.). Concord: House of Anansi Press.

Mejía, Rocío. (2002). "Globalization and Impacts on Mexican Women in the Context of Globalization." Gender focus IX AWID International Forum Reinventing Globalization, Guadalajara, México, October 2002. Online: <http://www.prd.org.mx/data/loadfile.php?id=6299&file=women.doc&tipo=application/msword>.

Mitra, Arup. (2008). "Social Capital, Livelihood and Upward Mobility." *Habitat International,* 32, 261–269.

Morley, David. (2000). *Home Territories: Media, Mobility and Identity.* New York: Routledge.

Morley, David, and Kevin Robins. (1995). *Spaces of Identity: Global Media, Electronic Landscapes and Cultural Boundaries.* New York: Routledge.

Nunn, Samuel. (2001). "Cities, Space, and the New World of Urban Law Enforcement Technologies." *Journal of Urban Affairs,* 23:3–4, .259–278.

Ong, Aihwa, and June Nash. (1994). *Spirits of Resistance and Capitalist Discipline: Factory Women in Malaysia.* New York: State University of New York Press.

Pacey, A. (1996). *The Culture of Technology.* Cambridge, MA: MIT Press.

Packer, Jeremy. (2003). "Disciplining Mobility: Governing and Safety." *Foucault, Cultural Studies, and Governmentality.* Jack Bratich, Jeremy Packer, and Cameron McCarthy (eds.). Albany: State University of New York Press, pp. 135–164.

Polan, Dana. (1993). "The Public's Fear: Or, Media as Monsters in Habermas, Negt, and Kluge." *The Phantom Public Sphere.* Bruce Robbins (ed.). Minneapolis: University of Minnesota Press, pp.33-41.

Postman, N. (1993). *Technopoly: The Surrender of Culture to Technology.* Toronto: Random House of Canada Ltd.

Puro, J. (2002). "Finland: A Mobile Culture." *Perpetual Contact.* J.E. Katz and M. Aakhus (Eds.). Cambridge: Cambridge University Press.

Rheingold, H. (2002). *Smart Mobs: The Next Social Revolution.* Cambridge, MA: Perseus.

Richardson, L. (2000a). "Writing: A Method of Inquiry." *The Sage Handbook of Qualitative Research.* N. K. Denzin and Y. K. Lincoln (eds.). Thousand Oaks, CA: Sage, pp.923–948.

———. (2000b). "Evaluating Ethnography." *Qualitative Inquiry,* 6:2, 253–255.

Richardson, L. and St. Pierre, E. A. (2005). "Writing: A Method of Inquiry." *The Sage Handbook of Qualitative Research,* 2nd edition. N. K. Denzin and Y. K. Lincoln (eds.). Thousand Oaks, CA: Sage, pp. 959–978

Robinson, Laura. (2007). "The Cyberself: The Self-ing Project Goes Online, Symbolic Interaction in the Digital Age." *New Media & Society,* 9:1, 93–110.

Ronai, C. (1999). "The Next Night Sous Rature: Wrestling with Derrida's Mimesis." *Qualitative Inquiry,* 5:1, 114–130.

Roy, Anjali Gera. (2007). "Digital Culture: Some Beings Who Exchange Information on the Internet." *The International Journal of Diversity in Organisations, Communities and Nations,* 7:2, pp.143–151.

Rutsky, R. L. (1999). *High Techné: Art and Technology from the Machine Aesthetic to the Posthuman.* Minneapolis: University of Minnesota Press.

Saldaña, Johnny. (2003). "Dramatizing Data: A Primer," *Qualitative Inquiry,* 9:2, 218–236.

Schick, Laurie. (2008). "Breaking Frame in a Role-Play Simulation: A Language Socialization Perspective." *Simulation & Gaming,* 39:2, pp.184–197.

Seale, Clive. (1999). "Quality in Qualitative Research." *Qualitative Inquiry,* 5:4, 465–478.

Seippel, Ørnulf. (2008). "Sports in Civil Society: Networks, Social Capital and Influence." *European Sociological Review,* 24:1, pp.69–80.

Simmons, Maxine. (2007). "Insider Ethnography: Tinker, Tailor, Researcher or Spy?" *Nurse Researcher,* 14:4, 7–17.

Soja, E. (1989). *Postmodern Geographies: The Reassertion of Space in Critical Social Theory.* London: Verso.

———. (1999). "History: Geography: Modernity." *The Cultural Studies Reader,* 2nd edition. Simon During (ed.). New York: Routledge, pp. 113–125.

Soukup, Charles. (2006). "Computer-Mediated Communication as a Virtual Third Place: Building Oldenburg's Great Good Places on the World Wide Web." *New Media & Society.* 8:3, 421–440.

Spigel, Lynn. (2001). "Media Homes: Then and Now." *International Journal of Cultural Studies,* 4:4, 385–411.

Star, Susan Leigh. (1991). "Power, Technologies and the Phenomenology of Conventions: On Being Allergic to Onions." *A Sociology of Monsters? Essays on Power, Technology and Domination, Sociological Review Monograph 38.* John Law (ed.). London: Routledge, pp.26–56.

Stivale, Charles. (1985). *Pragmatic/Machinic: A Discussion with Félix Guattari.* Online March 19, 1985: <http://www.dc.peachnet.edu/~mnunes/guattari.html#p2.3>.

Tettegah, Sharon, and Grant Kien. (2003a). "Identity, Human-Computer Interaction, and Teacher Voices." Paper presentation. *Conference of the American Educational Research Association (AERA),* Chicago, April 2003.

———. (2003b). "Animated Vignettes as an Online Education Tool." Paper presentation. *Conference of the Association of Internet Researchers – AOIR4.0* Toronto 2003: Broadening the Band, Toronto, Canada. October 2003.

Tomlinson, John. (1999). *Globalization and Culture.* Chicago: University of Chicago Press.

Ulmer, G. (1989). *Teletheory: Grammatology in the Age of Video.* New York: Routledge.

Urry, J. (2000). *Sociology beyond Societies: Mobilities for the Twenty-First Century.* New York: Routledge.

Vannini, Phillip. (2007). "Social Semiotics and Fieldwork: Method and Analytics." *Qualitative Inquiry,* 13, 113–140.

Vega, Iva'n De la, and Hebe Vessuri. (2008). "Science and Mobility: Is Physical Location Relevant?" *Technology in Society,* 30, 71–83.

Virilio, Paul. (1986). *Speed and Politics.* Mark Polizzotti (trans.). New York: Semiotext(e).

———. (1997). *Open Sky.* Julie Rose (trans.). New York: Verso.

———. (2000). *The Information Bomb.* Chris Turner (trans.). New York: Verso.

———. (2002). *Ground Zero.* Chris Turner (trans.). New York: Verso.

———. (2003). *Unknown Quantity.* New York: Thames & Hudson.

Virno, Paolo. (2004). *A Grammar of the Multitude: For an Analysis of Contemporary Forms of Life.* Isabella Bertoletti, James Cascaito, and Andrea Casson (trans.). New York: Semiotext(e).

Waskul, Dennis D. (2005). "Ekstasis and the Internet: Liminality and Computer-Mediated Communication." *New Media & Society,* 7:1, 47–63.

Whitty, Monica T. (2008). "Revealing the 'Real' Me, Searching for the 'Actual' You: Presentations of Self on an Internet Dating Site." *Computers in Human Behavior,* 24:4, 1707–1723.

Weiner, N. (1954). *The Human Use of Human Beings: Cybernetics and Society.* Boston: Da Capo Press.

Wilson, Brian. (2007). "New Media, Social Movements, and Global Sport Studies: A Revolutionary Moment and the Sociology of Sport." *Sociology of Sport Journal,* 24, pp.457–477.

Woodward, Ian, Zlatko Skrbis, and Clive Bean. (2008). "Attitudes towards Globalization and Cosmopolitanism: Cultural Diversity, Personal Consumption and the National Economy." *British Journal of Sociology,* 59:2, pp.207–226.

Xie, Shaobo. (2008). "Anxieties of Modernity: A Semiotic Analysis of Globalization Images in China." *Semiotica,* 170–1/4, pp.153–168.

Index

Actant, 4, 21, 56, 59–60, 64, 124, 150, 153, 169
Actor, 1, 4, 21, 28, 142, 169
  and power, 42, 50, 55
  and translation, 126
Actor-Network Theory (ANT), 3, 6, 18–19, 23, 54, 126, 169
Agency, 6, 29, 36–37, 39, 43, 51, 72, 77, 138, 149, 150, 165
Air force. *See* military
Alliance, ally, 19, 21, 60, 64, 87, 101, 127, 130, 137, 143, 170
Army. *See* military
Art, 4, 29, 152, 158, 162–165
Authentic, authenticity, 4, 7, 23, 31, 46, 82, 116, 141, 161, 166
Auto-mobility, automobile, 52, 56, 128

Baudrillard, Jean, 13, 29–30, 116, 127
Bauman, Zygmunt, 12, 13, 119, 152, 154
Beijing, China, 23, 142
Being, pure being, 13, 27, 57, 71, 72, 74, 76, 78, 79, 82, 141, 145, 151, 152, 155–162, 164
Belonging, 45, 52, 71, 80, 85, 87, 107, 113, 121, 124, 135, 155
Border, border crossing, 96, 119, 160

Callon, Michel, 3, 30, 37, 55
Canada, Canadian, 15, 44, 57–69, 85, 89, 105, 108, 121, 122, 130–145
Canadian Broadcasting Corporation (CBC), 85, 105, 108, 110, 118, 133
Carey, James, 3, 12, 16, 41–43, 58, 89, 119, 152
Castells, Manuel, 29, 36–37, 92, 99
Charland, Maurice, 59, 106, 132–133
Chiapas, Mexico, 2, 7, 90
Christians, Clifford G., 150
Citizen, citizenship, global citizen, 3, 4, 7, 52, 87, 92, 103, 119, 123, 128, 131
Communication, study of, 2, 41, 43, 58, 65, 104, 128, 129, 149
Consumption, 46, 62, 105, 117
Cultural practice, 7, 57, 68, 113, 122
  See also ritual
Cultural territory, 59, 108, 110, 112, 117, 130–135
Culture, 5, 7, 8, 10, 13, 58, 85, 86, 88, 93, 100, 106, 113, 116–117, 133, 138, 153
Cyborg, 144, 150

Dasein. *See* being, pure being
Democracy, 33, 104
Denzin, Norman K., 4, 6, 14, 19, 23, 82, 152, 166
Distanceclessness, 29, 128, 141, 142, 145, 165

Epiphany, 2, 4, 71
Ethnographic field, 3, 6, 9, 14, 16, 53, 54
Ethnography, 3, 5, 6, 9, 13, 19, 21, 54, 56, 153
Everyday life, 3, 5, 6, 9, 10, 13, 16, 21, 41, 42, 45, 54, 62, 78, 80, 99, 123, 138, 142, 167

Foucault, Michel, 41, 50, 52, 56, 65–66, 83, 104

Giddens, Anthony, 80, 126–127, 154
Global technography, 1, 5, 6, 9, 16

Globalization, 2, 29, 33, 36, 45, 52, 85, 86, 89, 92, 95, 99, 104, 110, 113, 116–117, 122, 135, 145
Goffman, Erving, 6, 15
Government, governmentality, 36, 39, 52, 56, 65, 100

Hay, James, 6, 49, 53, 54–56, 121, 129
Hegemony, 60, 109, 123, 130
See also power
Heidegger, Martin, 4, 5, 13, 15, 27, 29, 65, 72, 76, 78–80, 82, 128, 141, 144, 147
Home, 45, 71, 80, 81, 88, 119, 121, 123, 124, 126, 128–130

Identity, 4, 5, 49, 52, 54, 81, 87, 88, 107, 110, 111, 112–113, 116, 119, 127, 129, 134, 152
Innis, Harold, 62, 65
Italy, 1, 7

Language, 46, 78, 86, 157, 159
as culture, 116, 118
as power, 60, 109
Latour, Bruno, 3, 18, 19, 23, 33, 37, 50, 59–60, 129, 130, 150
Law, John, 3, 30, 33, 37, 55, 119, 124, 129
Leibniz, Gottfried, 5
Local, 7, 21, 29, 41, 45, 56, 59, 62, 89, 92, 104, 106, 110, 113, 117, 122, 134, 138
Love, 19, 139, 142

Materiality, 21, 55, 124
McCarthy, Cameron, 5, 85, 87, 92, 106, 116, 122, 133
McLuhan, Marshall, 14, 27, 29, 33, 58, 64, 90, 105, 154
Media (and media studies), 2, 4, 5, 20, 21, 33, 55, 59, 61, 64, 104, 126, 127, 132, 165, 166
Method, Actor-Network Theory, 6, 18, 21, 23, 173
Migration, 2, 45, 49, 87, 92, 95, 106
Military, 32, 34, 44, 47, 60, 90, 91, 96, 106
Mobile technography. *See* global technography
Mobility, 1–4, 6, 21, 29, 32, 33, 36, 41, 42, 45, 49, 51, 52, 55, 56, 57, 71, 85, 89, 92, 103, 106, 122, 124, 129, 135, 138, 139, 145, 152, 154
Montreal, Canada, 55, 108, 127, 131, 133
Morley, David, 45, 51, 107, 124, 126, 128

Nation, nationalism, 3, 4, 46, 49, 86, 87, 106–107, 110, 113, 122, 128, 130, 132, 134
Network, 1, 3, 4, 5, 6, 12, 21, 27, 29–33, 36–37, 41, 45, 49–50, 54–56, 60, 83, 99–100, 110, 119, 124, 128, 130, 132, 134, 145

Ontology, ontological security, 2, 5, 7, 27, 30, 32–33, 50, 71, 78, 79, 121, 126, 142, 149, 161

Panopticism, 41, 83
Performativity, 3, 49, 118, 132
Place. *See* local
Power, network power, 3, 5, 27, 29, 30, 32–33, 37, 41, 42, 45, 49, 50, 51, 55, 56, 60, 63, 104, 107, 126, 129, 132

Quebec City, Canada, 117

Race, racism, 47, 88, 89, 106, 112, 131, 132, 157
Ritual, 4, 23, 42, 49, 52, 58, 79, 80, 88–89, 122, 138, 152
See also cultural practice

Sacred, 80, 104, 141
Seoul, South Korea, 9, 22, 30, 46, 103
Society, 18, 36, 38–39, 56, 58, 99
Spatialization, 1, 29, 41, 42, 63, 86, 135
Subjectivity, 2, 7, 49, 82, 110, 119, 134, 138, 155, 161, 164

Technological nationalism, 132, 135
Tokyo, Japan, 12, 95, 97
Toronto, Canada, 59, 63, 71, 77, 119, 120, 124, 130
Translation, 4, 60, 65, 124, 126, 129

Vancouver, Canada, 85, 87, 128
Virilio, Paul, 12, 13, 32–34, 65, 99, 104, 106, 127, 150, 152, 154–155

War, 33, 51, 90, 106, 125
Wireless, 2, 13, 14, 21, 56, 119, 130, 135
Wireless mobility. *See* mobility

Zapatista movement, 2, 90, 104

Intersections in Communications and Culture

Global Approaches and Transdisciplinary Perspectives

General Editors: Cameron McCarthy & Angharad N. Valdivia

An Institute of Communications Research, University of Illinois Commemorative Series

This series aims to publish a range of new critical scholarship that seeks to engage and transcend the disciplinary isolationism and genre confinement that now characterizes so much of contemporary research in communication studies and related fields. The editors are particularly interested in manuscripts that address the broad intersections, movement, and hybrid trajectories that currently define the encounters between human groups in modern institutions and societies and the way these dynamic intersections are coded and represented in contemporary popular cultural forms and in the organization of knowledge. Works that emphasize methodological nuance, texture and dialogue across traditions and disciplines (communications, feminist studies, area and ethnic studies, arts, humanities, sciences, education, philosophy, etc.) and that engage the dynamics of variation, diversity and discontinuity in the local and international settings are strongly encouraged.

LIST OF TOPICS

- Multidisciplinary Media Studies
- Cultural Studies
- Gender, Race, & Class
- Postcolonialism
- Globalization
- Diaspora Studies
- Border Studies
- Popular Culture
- Art & Representation
- Body Politics
- Governing Practices
- Histories of the Present
- Health (Policy) Studies
- Space and Identity
- (Im)migration
- Global Ethnographies
- Public Intellectuals
- World Music
- Virtual Identity Studies
- Queer Theory
- Critical Multiculturalism

Manuscripts should be sent to:

Cameron McCarthy OR Angharad N. Valdivia
Institute of Communications Research
University of Illinois at Urbana-Champaign
222B Armory Bldg., 555 E. Armory Avenue
Champaign, IL 61820

FSC
www.fsc.org
MIX
Papier | Fördert gute Waldnutzung
FSC® C083411

Kolibri 360 GmbH
Ferdinand-Jühlke-Straße 7
99095 Erfurt, Deutschland
produktsicherheit@kolibri360.de